Springer Und

Springer
London
Berlin
Heidelberg
New York
Barcelona
Budapest
Hong Kong
Milan
Paris
Santa Clara
Singapore
Tokyo

Other books in this series

Basic Linear Algebra
T.S. Blyth and E.F. Robertson
3-540-76122-5

Elements of Logic via Numbers and Sets
D.L. Johnson
3-540-76123-3

Multivariate Calculus and Geometry
Seán Dineen
3-540-76176-4

Elementary Number Theory
Gareth A. Jones and J. Mary Jones
3-540-76197-7

Vector Calculus
P.C. Matthews
3-540-76180-2

Introductory Mathematics: Algebra and Analysis
Geoff Smith
3-540-76178-0

Gordon S. Marshall

Introductory Mathematics: Applications and Methods

With 43 Figures

Springer

Gordon S. Marshall, BSc, PhD
School of Computing and Mathematics, University of Teesside, Middlesbrough, Cleveland TS1 3BA, UK

Cover illustration elements reproduced by kind permission of:
Aptech Systems, Inc., Publishers of the GAUSS Mathematical and Statistical System, 23804 S.E. Kent-Kangley Road, Maple Valley, WA 98038, USA. Tel: (206) 432 - 7855 Fax (206) 432 - 7832 email: info@aptech.com URL: www.aptech.com
American Statistical Association: Chance Vol 8 No 1, 1995 article by KS and KW Heiner 'Tree Rings of the Northern Shawangunks' page 32 fig 2
Springer-Verlag: Mathematica in Education and Research Vol 4 Issue 3 1995 article by Roman E Maeder, Beatrice Amrhein and Oliver Gloor 'Illustrated Mathematics: Visualization of Mathematical Objects' page 9 fig 11, originally published as a CD ROM 'Illustrated Mathematics' by TELOS: ISBN 0-387-14222-3, german edition by Birkhauser: ISBN 3-7643-5100-4.
Mathematica in Education and Research Vol 4 Issue 3 1995 article by Richard J Gaylord and Kazume Nishidate 'Traffic Engineering with Cellular Automata' page 35 fig 2. Mathematica in Education and Research Vol 5 Issue 2 1996 article by Michael Trott 'The Implicitization of a Trefoil Knot' page 14.
Mathematica in Education and Research Vol 5 Issue 2 1996 article by Lee de Cola 'Coins, Trees, Bars and Bells: Simulation of the Binomial Process page 19 fig 3. Mathematica in Education and Research Vol 5 Issue 2 1996 article by Richard Gaylord and Kazume Nishidate 'Contagious Spreading' page 33 fig 1. Mathematica in Education and Research Vol 5 Issue 2 1996 article by Joe Buhler and Stan Wagon 'Secrets of the Madelung Constant' page50 fig 1.

ISBN 3-540-76179-9 Springer-Verlag Berlin Heidelberg New York

British Library Cataloguing in Publication Data
Marshall, G. S.
Introductory mathematics: applications and methods. -
(Springer undergraduate mathematics series)
1. Mathematics
I. Title
510
ISBN 3540761799

Library of Congress Cataloging-in-Publication Data
Marshall, G. S. (Gordon Stanley), 1961-
Introductory mathematics: applications and methods. /
G.S. Marshall.
p. cm. -- (Springer undergraduate mathematics series)
Includes index.
ISBN 3-540-76179-9 (pbk. : alk. paper)
1. Mathematics. I. Title. II. Series.
QA37.2.M3587 1998 97-40061
515--dc21 CIP

Typesetting: Camera ready by author
Printed and bound at the Athenæum Press Ltd., Gateshead, Tyne & Wear
12/3830-543210 Printed on acid-free paper

Preface

This book is aimed at undergraduate students embarking on the first year of a modular mathematics degree course. It is a self-contained textbook making it ideally suited to distance learning and a useful reference source for courses with the traditional lecture/tutorial structure.

The theoretical content is firmly based but the principal focus is on techniques and applications.

The important aims and objectives are presented clearly and then reinforced using complete worked solutions within the text. There is a natural increase in difficulty and understanding as each chapter progresses, always building upon the basic elements.

It is assumed that the reader has studied elementary calculus at Advanced level and is at least familiar with the concept of function and has been exposed to basic differentiation and integration techniques. Although these are covered in the book they are presented as a refresher course to jog the student's memory rather than to introduce the topic for the first time.

The early chapters cover the topics of matrix algebra, vector algebra and complex numbers in sufficient depth for the student to feel comfortable when they reappear later in the book.

Subsequent chapters then build upon the student's 'A' level knowledge in the area of real variable calculus, including partial differentiation and multiple integrals. The concluding chapter on differential equations motivates the student's learning by consideration of applications taken from both physical and economic contexts.

Mathematics is a practical subject which can never be mastered without attempting a significant number of examples. Throughout this book there are a large number of exercises of varying difficulty, some have been included with the sole intention of stretching the most able of students. Worked solutions are provided to all exercises.

Table of Contents

1. Simultaneous Linear Equations 1
- 1.1 Introduction 1
- 1.2 The method of determinants 2
- 1.3 Gaussian elimination 8
- 1.4 Ill–conditioning 10
- 1.5 Matrices 11
  - 1.5.1 Matrix notation 11
  - 1.5.2 Matrix algebra 12
  - 1.5.3 Inverse matrices 17
  - 1.5.4 Matrix solution of linear equations 19

2. Vector Algebra 21
- 2.1 Introduction 21
- 2.2 Algebraic representation 22
- 2.3 Linear independence 24
- 2.4 The scalar product 25
- 2.5 The vector product 26
- 2.6 Triple products 28
- 2.7 Differentiation of vector functions 31

3. Complex Numbers 33
- 3.1 Introduction 33
- 3.2 Algebra of complex numbers 34
- 3.3 Graphical representation 35
- 3.4 Polar form 36
- 3.5 Exponential form 38
- 3.6 De Moivre's theorem 40
  - 3.6.1 Roots of unity 41
  - 3.6.2 Roots of complex numbers 42

4. **Review of Differentiation Techniques** ... 43
4.1 Introduction ... 43
4.2 Differentiation of standard functions ... 45
4.3 Function of a function ... 47
4.4 The product rule ... 49
4.5 The quotient rule ... 51
4.6 Higher derivatives ... 53
4.7 Implicit differentiation ... 54
4.8 Logarithmic differentiation ... 55

5. **Review of Integration Techniques** ... 57
5.1 Introduction ... 57
5.2 Integration of standard functions ... 60
5.3 Integration by substitution ... 62
5.4 Integration using partial fractions ... 67
5.4.1 Introduction ... 67
5.4.2 Resolving into partial fractions ... 70
5.4.3 Repeated factors ... 73
5.5 Integration by parts ... 76

6. **Applications of Differentiation** ... 79
6.1 Introduction ... 79
6.2 Functions ... 79
6.2.1 Increasing and decreasing functions ... 79
6.2.2 Relative maxima and minima ... 80
6.2.3 Concavity ... 82
6.2.4 Points of inflection ... 84
6.2.5 Absolute maxima and minima ... 84
6.3 Curve sketching ... 91
6.4 Optimisation ... 94
6.5 Taylor series ... 97

7. **Partial Differentiation** ... 103
7.1 Introduction ... 103
7.2 Notation ... 103
7.3 Total differentials ... 108
7.4 The total derivative ... 110
7.5 Taylor series for functions of two variables ... 112
7.6 Maxima, minima and saddlepoints ... 115
7.6.1 Classification of critical points ... 116
7.7 Problems with constraints ... 119
7.8 Lagrange multipliers ... 122

8. Multiple Integrals 127
8.1 Double integrals 127
8.1.1 Change of variable in double integrals 132
8.2 Triple integrals 134

9. Differential Equations 137
9.1 Introduction 137
9.2 First order differential equations 142
9.2.1 Equations of the form $\frac{dy}{dx} = f(x)$ 142
9.2.2 Equations of the form mr$\frac{dy}{dx} = ky$ 151
9.2.3 Variables separable 157
9.2.4 Integrating factors 163
9.3 Second order equations 168
9.3.1 Linear equations with constant coefficients 168
9.3.2 Solution of the homogenous equation 169
9.3.3 Simple harmonic motion 172
9.3.4 Damped SHM 175
9.3.5 Particular integrals involving elementary functions 178

Solutions to Exercises 185

Index 225

1
Simultaneous Linear Equations

1.1 Introduction

Systems of simultaneous linear equations are commonplace in optimisation problems, statistics, economics and process control. The following example illustrates a simple application of linear equations.

A small business has £7200 available to update its personal computers. New machines are available at £1500 each or second-hand machines could be bought for £700 each. How many of each should be bought if the business needs 8 machines and spends all the available money?

This problem can be represented as a pair of linear equations by letting N denote the number of new machines to be bought and S the number of second-hand machines.

$$\begin{aligned} 1500N + 700S &= 7200 \\ N + S &= 8 \end{aligned}$$

This example and its solution appears as one of the exercises in this chapter.

In this chapter various methods of solving simultaneous linear equations are introduced, each method having advantages and limitations. First of all the method of determinants is considered for small systems of equations (this work is useful for introducing ideas that are met again in the sections on matrices and in the chapters on vector algebra and multiple integrals), followed by Gaussian elimination and finally the use of matrices.

1.2 The method of determinants

Suppose that we wish to solve the following pair of equations for x and y

$$\begin{aligned} ax + by &= c \\ lx + my &= n \end{aligned}$$

where a, b, c, l, m and n are constants.

Multiplying the first equation by m, the second equation by b and subtracting eliminates y leaving us with the linear equation in x

$$(am - lb)x = cm - nb\ .$$

Similarly, multiplying the first equation by l, the second by a and subtracting would eliminate x and give us a linear equation in y only

$$(bl - am)y = cl - na\ .$$

The solution to the pair of equations is then

$$x = \frac{cm - nb}{am - lb}\ , \qquad y = \frac{an - lc}{am - lb}$$

provided $am - lb \neq 0$. The denominator in these expressions for x and y is called the *determinant* of the coefficients of x and y and is defined by the notation

$$\begin{vmatrix} a & b \\ l & m \end{vmatrix} = am - lb\ .$$

This is called a second order determinant because it has 2 rows and 2 columns bounded by two *vertical lines*, not brackets. A third order determinant would have 3 rows and 3 columns. A determinant has a single value, for example

$$\begin{vmatrix} 2 & 4 \\ 1 & 3 \end{vmatrix} = 2 \times 3 - 4 \times 1 = 5\ .$$

Notice now that x and y can be written as fractions involving two determinants

$$x = \frac{\begin{vmatrix} c & b \\ n & m \end{vmatrix}}{\begin{vmatrix} a & b \\ l & m \end{vmatrix}}\ , \qquad y = \frac{\begin{vmatrix} a & c \\ l & n \end{vmatrix}}{\begin{vmatrix} a & b \\ l & m \end{vmatrix}}\ .$$

The determinants in the numerators are very similar to the denominator. We write the solution in the form $x = \frac{\Delta_x}{\Delta}$, $\quad y = \frac{\Delta_y}{\Delta}$ where

$$\Delta - \begin{vmatrix} a & b \\ l & m \end{vmatrix}, \quad \Delta_x = \begin{vmatrix} c & b \\ n & m \end{vmatrix}, \quad \Delta_y = \begin{vmatrix} a & c \\ l & n \end{vmatrix}$$

The determinants Δ_x and Δ_y are obtained from Δ by replacing the first and second columns respectively by the right-hand sides of the original linear equations.

Remember, the solution is defined only if $\Delta \neq 0$. This approach is called the method of determinants, or Cramer's rule.

Problem 1.1

Use the method of determinants to solve the following pair of equations

$$\begin{aligned} 5x - 2y &= 1 \\ 3x + 4y &= 2\,. \end{aligned}$$

Solution 1.1

We first calculate the three determinants:

$$\begin{aligned} \Delta &= \begin{vmatrix} 5 & -2 \\ 3 & 4 \end{vmatrix} = 5 \times 4 - (-2) \times 3 = 26 \\ \Delta_x &= \begin{vmatrix} 1 & -2 \\ 2 & 4 \end{vmatrix} = 1 \times 4 - (-2) \times 2 = 8 \\ \Delta_y &= \begin{vmatrix} 5 & 1 \\ 3 & 2 \end{vmatrix} = 5 \times 2 - 1 \times 3 = 7\,. \end{aligned}$$

Hence the solution is $x = \frac{\Delta_x}{\Delta} = \frac{8}{26}$, $y = \frac{\Delta_y}{\Delta} = \frac{7}{26}$.
Check the result by substituting into the original equations:

$$\begin{aligned} 5x - 2y &= 5 \times \frac{8}{26} - 2 \times \frac{7}{26} = \frac{26}{26} = 1 \\ 3x + 4y &= 3 \times \frac{8}{26} + 4 \times \frac{7}{26} = \frac{52}{26} = 2\,. \end{aligned}$$

It is rare for the coefficients to be integers. Often the linear equations are obtained from measurement and so given to a number of decimal places.

Problem 1.2

Use Cramer's rule to solve for x and y to 3 decimal place accuracy

$$\begin{aligned} 1.46x + 3.12y &= 3.04 \\ 2.18x - 1.69y &= -3.17\,. \end{aligned}$$

Solution 1.2

To obtain solutions accurate to 3 decimal places we must keep at least 4 decimal places during the intermediate stages. The rounding to 3 decimal places is the *final* step. The three determinants are

$$\begin{aligned}
\Delta &= \begin{vmatrix} 1.46 & 3.12 \\ 2.18 & -1.69 \end{vmatrix} = -2.4674 - 6.8016 = -9.2690 \\
\Delta_x &= \begin{vmatrix} 3.04 & 3.12 \\ -3.17 & -1.69 \end{vmatrix} = -5.1376 - (-9.8904) = 4.7528 \\
\Delta_y &= \begin{vmatrix} 1.46 & 3.04 \\ 2.18 & -3.17 \end{vmatrix} = -4.6282 - 6.6272 = -11.2554\,.
\end{aligned}$$

Hence the values of x and y are

$$x = \frac{\Delta_x}{\Delta} = \frac{4.7528}{-9.2690} = -0.5128,\ y = \frac{\Delta_y}{\Delta} = \frac{-11.2554}{-9.2690} = 1.2143\,.$$

The solution (to 3 d.p.) is $x = -0.513$ and $y = 1.214$.
Check this solution by substituting into the original equations.

$$\begin{aligned}
1.46x + 3.12y &= -0.74898 + 3.78768 = 3.03870 = 3.04 \text{ 2d.p.} \\
2.18x - 1.69y &= -1.11834 - 2.05166 = -3.17000 = -3.17 \text{ 2d.p.}
\end{aligned}$$

The method of determinants is straightforward for a pair of linear equations. *Can it be extended to larger systems?* The answer is yes but the effort involved increases rapidly with more equations. Suppose we have three linear simultaneous equations in x, y and z

$$\begin{aligned}
ax + by + cz &= d \\
ex + fy + gz &= h \\
jx + ky + mz &= n
\end{aligned}$$

where the symbols a to n are constants. (Later we introduce a subscript notation for defining large numbers of constants which is much neater.)

If we extend our earlier ideas we logically get

$$\Delta = \begin{vmatrix} a & b & c \\ e & f & g \\ j & k & m \end{vmatrix} \qquad \Delta_x = \begin{vmatrix} d & b & c \\ h & f & g \\ n & k & m \end{vmatrix}$$

$$\Delta_y = \begin{vmatrix} a & d & c \\ e & h & g \\ j & n & m \end{vmatrix} \qquad \Delta_z = \begin{vmatrix} a & b & d \\ e & f & h \\ j & k & n \end{vmatrix}$$

and then the solution is

$$x = \frac{\Delta_x}{\Delta}\,, \; y = \frac{\Delta_y}{\Delta}\,, \; z = \frac{\Delta_z}{\Delta}\,.$$

The determinants Δ are now said to be third order (three rows and three columns) but how do we calculate their value?

If we return to our system of linear equations and solve them by elimination it is possible to get the following expression for x

$$x = \frac{d(fm - gk) - h(bm - ck) + n(bg - cf)}{a(fm - gk) - e(bm - ck) + j(bg - cf)}\,.$$

The bracketed terms are second order determinants using the elements of the second and third columns of the determinant Δ. The only difference between the numerator and denominator are the multipliers of these brackets. In the denominator it is the original x coefficients while in the numerator it is the right hand sides of the linear equations. Comparing the denominator in the solution for x with Δ we see that a third order determinant can be written as three second order determinants

$$\Delta = \begin{vmatrix} a & b & c \\ e & f & g \\ j & k & m \end{vmatrix} = a\begin{vmatrix} f & g \\ k & m \end{vmatrix} - e\begin{vmatrix} b & c \\ k & m \end{vmatrix} + j\begin{vmatrix} b & c \\ f & g \end{vmatrix}$$

with similar expressions for Δ_x, Δ_y and Δ_z.

The expressions in the numerator and denominator for x can be factorised in a number of different ways, for example

$$\begin{aligned} & a(fm - gk) - e(bm - ck) + j(bg - cf) \\ = \; & -b(em - gj) + f(am - cj) - k(ag - ce) \\ = \; & c(ek - fj) - g(ak - bj) + m(af - be)\,. \end{aligned}$$

These expressions represent the expansion of the determinant using the first, second and third columns respectively. We can expand the determinant using *any row or any column* – the result will be exactly the same. The only detail we have to be careful about is the minus signs on the multiplying coefficients.

Using the first column resulted in only one minus sign, $-e$, while expanding about the second column resulted in two minus signs, $-b$ and $-k$. These follow the checkerboard pattern

$$\begin{vmatrix} + & - & + \\ - & + & - \\ + & - & + \end{vmatrix}\,.$$

Problem 1.3

Solve the following three linear equations using determinants.

$$\begin{aligned} 3x + 2y + z &= 4 \\ x - y + 2z &= -7 \\ 2x + 3y + 5z &= -7 \end{aligned}$$

Solution 1.3

We calculate the four determinants, expanding about the first column.

$$\begin{aligned}
\Delta &= \begin{vmatrix} 3 & 2 & 1 \\ 1 & -1 & 2 \\ 2 & 3 & 5 \end{vmatrix} = 3\begin{vmatrix} -1 & 2 \\ 3 & 5 \end{vmatrix} - 1\begin{vmatrix} 2 & 1 \\ 3 & 5 \end{vmatrix} + 2\begin{vmatrix} 2 & 1 \\ -1 & 2 \end{vmatrix} \\
&= -33 - 7 + 10 = -30 \\
\Delta_x &= \begin{vmatrix} 4 & 2 & 1 \\ -7 & -1 & 2 \\ -7 & 3 & 5 \end{vmatrix} = 4\begin{vmatrix} -1 & 2 \\ 3 & 5 \end{vmatrix} - (-7)\begin{vmatrix} 2 & 1 \\ 3 & 5 \end{vmatrix} + (-7)\begin{vmatrix} 2 & 1 \\ -1 & 2 \end{vmatrix} \\
&= -44 + 49 - 35 = -30 \\
\Delta_y &= \begin{vmatrix} 3 & 4 & 1 \\ 1 & -7 & 2 \\ 2 & -7 & 5 \end{vmatrix} = 3\begin{vmatrix} -7 & 2 \\ -7 & 5 \end{vmatrix} - 1\begin{vmatrix} 4 & 1 \\ -7 & 5 \end{vmatrix} + 2\begin{vmatrix} 4 & 1 \\ -7 & 2 \end{vmatrix} \\
&= -63 - 27 + 30 = -60 \\
\Delta_z &= \begin{vmatrix} 3 & 2 & 4 \\ 1 & -1 & -7 \\ 2 & 3 & -7 \end{vmatrix} = 3\begin{vmatrix} -1 & -7 \\ 3 & -7 \end{vmatrix} - 1\begin{vmatrix} 2 & 4 \\ 3 & -7 \end{vmatrix} + 2\begin{vmatrix} 2 & 4 \\ -1 & -7 \end{vmatrix} \\
&= 84 + 26 - 20 = 90
\end{aligned}$$

and hence the solution is

$$x = \frac{\Delta_x}{\Delta} = 1, \quad y = \frac{\Delta_y}{\Delta} = 2, \quad z = \frac{\Delta_z}{\Delta} = -3\,.$$

The method of determinants is a straightforward means of solving small systems of linear equations. For larger systems of equations the determinants become more involved and the effort required increases rapidly. For example, a fourth order determinant would reduce to 4 third order determinants each of which reduces to 3 second order determinants. Although the method does not get any more difficult the increased number of calculations makes mistakes more likely when the work is done by hand.

Properties of determinants

There are properties of determinants which can be used to simplify the work.

1. Adding a multiple of a row to another row does not change the value of the determinant.
2. Adding a multiple of a column to another column does not change the value of the determinant.

We illustrate this using second order determinants but the manipulations apply to any order of determinant. For example, if k is a scalar then adding k times the second row to the top row does not change the value of the determinant. It is left as an exercise to verify that the same holds for columns.

$$\begin{aligned} \begin{vmatrix} a+kl & b+km \\ l & m \end{vmatrix} &= m(a+kl) - l(b+km) \\ &= am + mkl - bl - lkm \\ &= am - bl = \begin{vmatrix} a & b \\ l & m \end{vmatrix}. \end{aligned}$$

EXERCISES

Solve the following pairs of simultaneous linear equations using the method of determinants (Cramer's rule). Answer to 3 decimal place accuracy where necessary and apply a check in each case.

1.1. $\begin{aligned} 3x - 7y &= 47 \\ 5x + 2y &= 10 \end{aligned}$

1.2. $\begin{aligned} 1.985x - 1.358y &= 2.212 \\ 0.953x - 0.652y &= 1.062 \end{aligned}$

1.3. Use determinants to show that the following pairs of linear equations do not have a unique solution. If the solution to a pair of linear equations represents the point of intersection of two straight lines, give a geometric explanation for the lack of a unique solution.

(a) $\begin{aligned} 2x + 3y &= 4 \\ 4x + 6y &= 5 \end{aligned}$ (b) $\begin{aligned} x - 2y &= 3 \\ 3x - 6y &= 9 \end{aligned}$

1.4. Solve the following system of linear equations using the method of determinants.

$$\begin{aligned} x + 3y + z &= 3 \\ 2x + y + 4z &= -1 \\ 3x + y - 2z &= 6 \end{aligned}$$

1.3 Gaussian elimination

The method of determinants is an efficient method for small systems of linear equations but it is inappropriate for larger systems. This section describes the method known as Gaussian elimination.

When using Gaussian elimination the objective is to transform the original system into an *upper triangular* form with non-zero diagonal entries. Suppose we begin with a system of three equations in the three unknowns x_1, x_2 and x_3 (this general notation makes it easy to extend the method to larger systems of linear equations),

$$\begin{aligned} a_{11}x_1 + a_{12}x_2 + a_{13}x_3 &= r_1 \qquad (1) \\ a_{21}x_1 + a_{22}x_2 + a_{23}x_3 &= r_2 \qquad (2) \\ a_{31}x_1 + a_{32}x_2 + a_{33}x_3 &= r_3 \qquad (3) \end{aligned}$$

where the coefficients a_{ij} and r_i are constants. The form we seek is

$$\begin{array}{llll} b_{11}x_1 & +b_{12}x_2 & +b_{13}x_3 & = s_1 \\ & b_{22}x_2 & +b_{23}x_3 & = s_2 \\ & & b_{33}x_3 & = s_3 \ . \end{array}$$

If we can achieve this then the variables x_1, x_2, x_3 are easily found using back substitution, starting with x_3. There are two stages of elimination. Stage 1 is to eliminate x_1 from equations (2) and (3), using equation (1). Then we eliminate x_2 from the third equation to give the upper triangular form.

Eliminate x_1 from (2) and (3) using

$$\text{equation}(i) - \frac{a_{i1}}{a_{11}} \times \text{equation}(1)$$

where $i = 2, 3$. We now have

$$\begin{array}{lllll} a_{11}x_1 & +a_{12}x_2 & +a_{13}x_3 & = r_1 & (1) \\ & b_{22}x_2 & +b_{23}x_3 & = s_2 & (4) \\ & b_{32}x_2 & +b_{33}x_3 & = s_3 & (5) \ . \end{array}$$

Now we eliminate x_2 from equation (5) using

$$\text{equation}(5) - \frac{b_{32}}{b_{22}} \times \text{equation}(4) \ .$$

If there were n such linear equations then we would eliminate x_1 from $(n-1)$ equations, then eliminate x_2 from $(n-2)$ equations, then eliminate x_3 from $(n-3)$ equations and so on. The advantage of this method is it can be made algorithmic and thus easy to implement as a computer program.

Problem 1.4

Solve the following three linear equations using Gaussian elimination

$$\begin{array}{rcrcrcrl} 3x & + & 2y & + & z & = & 4 & \quad (1) \\ x & - & y & + & 2z & = & -7 & \quad (2) \\ 2x & + & 3y & + & 5z & = & -7 & \quad (3)\,. \end{array}$$

Solution 1.4

Eliminate x from (2) and (3) using

$$\begin{array}{lll} \text{equ}(2) - \dfrac{a_{21}}{a_{11}} \times \text{equ}(1) & \text{i.e.} & \text{equ}(2) - \dfrac{1}{3} \times \text{equ}(1) \Rightarrow -\dfrac{5}{3}y + \dfrac{5}{3}z = -\dfrac{25}{3} \\ \text{equ}(3) - \dfrac{a_{31}}{a_{11}} \times \text{equ}(1) & \text{i.e.} & \text{equ}(3) - \dfrac{2}{3} \times \text{equ}(1) \Rightarrow \dfrac{5}{3}y + \dfrac{13}{3}z = -\dfrac{29}{3} \end{array}$$

Clearing the fractions we have

$$\begin{array}{rcrcrcrl} 3x & + & 2y & + & z & = & 4 & \quad (1) \\ & & y & - & z & = & 5 & \quad (4) \\ & & 5y & + & 13z & = & -29 & \quad (5)\,. \end{array}$$

Now we eliminate y from equation (5) using

$$\text{equation}(5) - \frac{5}{1} \times \text{equation}(4) \rightarrow 18z = -54$$

to give the triangular form we require,

$$\begin{array}{rcrcrcrl} 3x & + & 2y & + & z & = & 4 & \quad (1) \\ & & y & - & z & = & 5 & \quad (4) \\ & & & & 18z & = & -54 & \quad (6)\,. \end{array}$$

Back substitution then gives $z = -3$, $y = 2$, $x = 1$.

EXERCISES

Use Gaussian elimination to solve the following sets of linear equations.

1.5.

$$\begin{aligned} x + 3y + z &= 3 \\ 2x + y + 4z &= -1 \\ 3x + y - 2z &= 6 \end{aligned}$$

1.6.

$$\begin{aligned} 2x + y + 4z &= 17 \\ 3x - 3y - z &= 0 \\ x + z &= 5 \end{aligned}$$

1.4 Ill–conditioning

The solution to the pair of equations

$$\begin{aligned} 1.985x - 1.358y &= 2.212 \\ 0.953x - 0.652y &= 1.062 \end{aligned}$$

is $x = 0.609$ and $y = -0.739$, to three decimal places, see exercise 1.2.

Suppose we change the right hand side of the second equation only slightly, so that 1.062 becomes 1.061, i.e.

$$\begin{aligned} 1.985x - 1.358y &= 2.212 \\ 0.953x - 0.652y &= 1.061 , \end{aligned}$$

then intuitively we would expect a small change in the solution. However, in this case the solution is $x = 30.01$ and $y = 42.41$.

A small change in one coefficient has altered the solution drastically. These equations are said to be *ill–conditioned.*

The problem is that we are looking for the point of intersection of two lines which are very nearly parallel. As the coefficients have been rounded, each equation is represented by a band rather than a straight line.

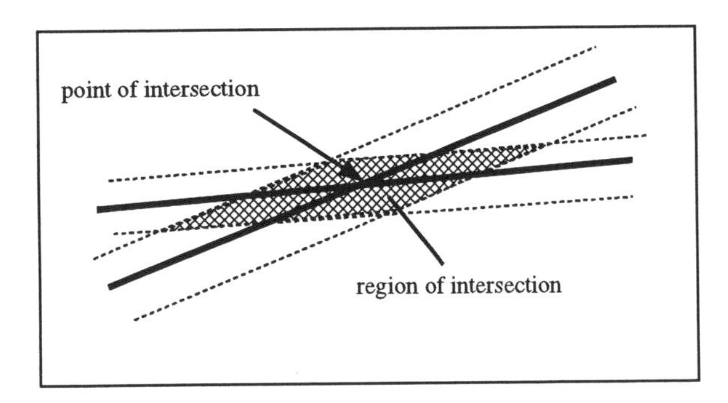

Fig. 1.1. Ill–conditioning

As the lines become closer to being parallel the region of possible intersection of the two bands, hatched in figure 1.1, becomes longer and changing the coefficient in one equation by a small amount may change the solution by a disproportionately large amount.

Numerically, the problem is that the determinant of the coefficient matrix is very small (almost singular), in this example it is approximately -0.000046, and round-off errors become very important. Fortunately, ill–conditioning rarely occurs; nevertheless we should be wary when we get very small values of determinants.

1.5 Matrices

1.5.1 Matrix notation

A matrix is a rectangular array of *elements* within a pair of brackets, the elements usually being real numbers. The most important thing to remember is that a matrix has ROWS and COLUMNS, in that order. If there are m *rows* and n *columns* then the matrix is said to have *order* $m \times n$ (read as 'm by n').

For example, a matrix of order 2×3 has 2 rows and 3 columns, such as

$$\begin{pmatrix} 3 & -2 & 6 \\ -1 & 0 & 11 \end{pmatrix} \qquad \begin{pmatrix} a & b & c \\ d & e & f \end{pmatrix}$$

A *square* matrix is one which has the same number of rows and columns, $m = n$.

$$\begin{pmatrix} 1 & 0 \\ 0 & 2 \end{pmatrix} \qquad \begin{pmatrix} 3 & -2 & 0 \\ -1 & 11 & 0 \\ 0 & 6 & 1 \end{pmatrix}$$

To avoid writing out the matrix in full we use capital letters as a shorthand.

$$A = \begin{pmatrix} 3 & -2 & 6 \\ -1 & 0 & 11 \end{pmatrix} \quad B = \begin{pmatrix} 3 & -2 & 0 \\ -1 & 11 & 0 \\ 0 & 6 & 1 \end{pmatrix}$$

We use the lower case letter to indicate specific elements, thus a_{ij} is the element in row i and column j of A. Rows are numbered from top to bottom and columns from left to right. For the above matrices we have $a_{23} = 11$ and $b_{23} = 0$. The negative of a matrix A is denoted $-A$ and its elements are $-a_{ij}$. For the above

$$-A = \begin{pmatrix} -3 & 2 & -6 \\ 1 & 0 & -11 \end{pmatrix} \qquad -B = \begin{pmatrix} -3 & 2 & 0 \\ 1 & -11 & 0 \\ 0 & -6 & -1 \end{pmatrix}.$$

Some frequently occurring types of matrices are given specific names.

- The *diagonal* elements of a square matrix are a_{ii}, i.e. a_{11}, a_{22}, a_{33}, etc. A square matrix is *diagonal* if all the elements not on the diagonal are zero.
- A matrix with only one column is called a *column vector*, its order is $m \times 1$.
- A matrix with only one row is called a *row vector*, its order is $1 \times n$.

Row and column vectors appear when we use inverse matrices to solve systems of linear equations and also in the next chapter on vector algebra.

The *transpose* of a matrix is obtained by interchanging the rows and columns and is denoted by A^T. This means that the transpose of an $m \times n$ matrix has order $n \times m$ and element a_{ij} of A becomes element a_{ji} of A^T. For example,

$$A = \begin{pmatrix} 3 & -2 & 6 \\ -1 & 0 & 11 \end{pmatrix}, \quad A^T = \begin{pmatrix} 3 & -1 \\ -2 & 0 \\ 6 & 11 \end{pmatrix}.$$

If a matrix and its transpose are equal, $A = A^T$, and the matrix is said to be *symmetric*. Note that A must be square for A and its transpose A^T to have the same order.

EXERCISES

Given the following matrices, answer the subsequent questions.

$$A = \begin{pmatrix} 3 & -2 & 6 \\ -1 & 0 & 11 \end{pmatrix}, \quad B = \begin{pmatrix} 3 & -2 & 0 \\ -1 & 11 & 0 \\ 0 & 6 & 1 \end{pmatrix}, \quad C = \begin{pmatrix} 4 & -3 & 1 \end{pmatrix}$$

1.7. What is the order of each of the above matrices?

1.8. Write down the elements a_{11}, b_{32} and c_{13}.

1.9. Calculate $a_{12} \times b_{33}$ and $a_{22} \times c_{12}$.

1.5.2 Matrix algebra

The rules for addition, subtraction and multiplication of numbers in arithmetic can be applied, with slight modifications, to matrices.

Equality of matrices

Two matrices are said to be equal if the matrices have the same order and the corresponding elements are equal. Solving the matrix equation

$$\begin{pmatrix} 3 & -2 & 6 \\ -1 & 0 & 11 \end{pmatrix} = \begin{pmatrix} a & -2 & 6 \\ -1 & e & f \end{pmatrix}$$

by equating the elements, we see that $a = 3$, $e = 0$ and $f = 11$.

Problem 1.5

Find the values of x and y satisfying the following matrix equation

$$\begin{pmatrix} 0 & 2 \\ 4 & 4 \end{pmatrix} = \begin{pmatrix} y+x & z+1 \\ y-x & x^2 \end{pmatrix}.$$

Solution 1.5

Equating the individual elements we have four equations

$$y + x = 0 \qquad z + 1 = 2$$
$$y - x = 4 \qquad x^2 = 4\,.$$

The equation involving z gives $z = 1$ and the pair of linear equations in x and y has the solution $x = -2$, $y = 2$. Finally, $x^2 = 4 \Rightarrow x = \pm 2$ but we have already found that only $x = -2$ satisfies the linear equations.

Addition and subtraction

To add or subtract two matrices they must have the same order. Given this we then add or subtract corresponding elements. For example

$$\begin{pmatrix} 3 & -2 \\ 6 & 8 \\ -1 & 0 \end{pmatrix} + \begin{pmatrix} 4 & -3 \\ 5 & -2 \\ 0 & 7 \end{pmatrix} = \begin{pmatrix} 3+4 & -2+(-3) \\ 6+5 & 8+(-2) \\ -1+0 & 0+7 \end{pmatrix} = \begin{pmatrix} 7 & -5 \\ 11 & 6 \\ -1 & 7 \end{pmatrix}$$

$$\begin{pmatrix} 3 & -2 \\ 6 & 8 \\ -1 & 0 \end{pmatrix} - \begin{pmatrix} 4 & -3 \\ 5 & -2 \\ 0 & 7 \end{pmatrix} = \begin{pmatrix} 3-4 & -2-(-3) \\ 6-5 & 8-(-2) \\ -1-0 & 0-7 \end{pmatrix} = \begin{pmatrix} -1 & 1 \\ 1 & 10 \\ -1 & -7 \end{pmatrix}.$$

Scalar multiplication

When a matrix is multiplied by a number this is referred to as scalar multiplication. Each element of the original matrix is multiplied by the scalar.

$$3\begin{pmatrix} 3 & -2 & 6 \\ -1 & 0 & 11 \end{pmatrix} = \begin{pmatrix} 3\times 3 & 3\times(-2) & 3\times 6 \\ 3\times(-1) & 3\times 0 & 3\times 11 \end{pmatrix} = \begin{pmatrix} 9 & -6 & 18 \\ -3 & 0 & 33 \end{pmatrix}$$

Matrix multiplication

It is not always possible to multiply two matrices together. The order of multiplication is important. As we shall see shortly, in general $AB \neq BA$, i.e. matrix multiplication is not commutative.

Matrix multiplication involves multiplying each *row* of the first matrix by each *column* of the second matrix. (Remember! *rows then columns.*) For this reason the number of columns in the first matrix must be equal to the number of rows in the second. If the number of columns of A is not equal to the number of rows of B then the product AB is not defined.

The easiest way to check that the product of two matrices is defined is to write down the order of each matrix (in the order of multiplication), and if the middle pair of numbers is identical then multiplication is possible. The outer pair of numbers then gives the order of the resulting product.

Problem 1.6

Given the matrices

$$A = \begin{pmatrix} 3 & -2 & 6 \\ -1 & 0 & 11 \end{pmatrix}, \quad B = \begin{pmatrix} 3 & -2 & 0 \\ -1 & 11 & 0 \\ 0 & 6 & 1 \end{pmatrix}, \quad C = \begin{pmatrix} 4 & -3 & 1 \end{pmatrix}$$

which of the following are defined? (a) AB, (b) BA, (c) AC, (d) CB. For those that *are* defined write down the order of the product.

Solution 1.6

The orders of the matrices A, B, C are 2×3, 3×3, 1×3 respectively. Write down the orders of the matrices as they appear in the products:

(a) $(2 \times \underline{3})\ (\underline{3} \times 3)$ the inner numbers match, therefore the product exists and its order is 2×3, the outer numbers.
(b) $(3 \times \underline{3})\ (\underline{2} \times 3)$ the inner numbers are different and therefore the product BA is not defined.
(c) $(2 \times \underline{3})\ (\underline{1} \times 3)$ the inner numbers are different and therefore the product AC is not defined.
(d) $(1 \times \underline{3})\ (\underline{3} \times 3)$ the inner numbers match, therefore the product exists and its order is 1×3, the outer numbers.

The next stage is to find each element of the product. Suppose A is order $m \times n$ and B is order $n \times p$ then the product AB exists and its order is $m \times p$. Let $C = AB$, then

$$\begin{pmatrix} a_{11} & \dots & a_{1n} \\ \vdots & \vdots & \vdots \\ a_{i1} & \dots & a_{in} \\ \vdots & \vdots & \vdots \\ a_{m1} & \dots & a_{mn} \end{pmatrix} \begin{pmatrix} b_{11} \dots & b_{1j} & \dots b_{1p} \\ \vdots & \vdots & \vdots \\ b_{n1} \dots & b_{nj} & \dots b_{np} \end{pmatrix} = \begin{pmatrix} c_{11} & \dots & c_{1p} \\ \vdots & \vdots & \vdots \\ \dots & c_{ij} & \dots \\ \vdots & \vdots & \vdots \\ c_{m1} & \dots & c_{mp} \end{pmatrix}$$

where

$$c_{ij} = a_{i1}b_{1j} + a_{i2}b_{2j} + a_{i3}b_{3j} + \ldots + a_{in}b_{nj}\ .$$

Thus the element c_{ij} is obtained by multiplying the first element of row i in A by the first element of column j in B then adding the product of the second element of row i in A by the second element of column j in B and so on.

Although this expression may look cumbersome, once we have tried a few examples the method becomes clearer.

Problem 1.7

Given the matrices

$$A = \begin{pmatrix} 3 & -2 & 6 \\ -1 & 0 & 11 \end{pmatrix}, \quad B = \begin{pmatrix} 3 & -2 & 0 \\ -1 & 11 & 0 \\ 0 & 6 & 1 \end{pmatrix}, \quad C = \begin{pmatrix} 4 & -3 & 1 \end{pmatrix}$$

find the products $D = CB$ and $E = AB$.

Solution 1.7

We have already found in problem 1.6 that both products are defined and their orders are 1×3 and 2×3 respectively. Consider $D = CB$ first.

$$D = CB = \begin{pmatrix} 4 & -3 & 1 \end{pmatrix} \begin{pmatrix} 3 & -2 & 0 \\ -1 & 11 & 0 \\ 0 & 6 & 1 \end{pmatrix}$$

The three elements of the product d_{11}, d_{12}, d_{13} are determined by multiplying the only row of C by each column of B in turn.

$$\begin{aligned} d_{11} &= 4 \times 3 + (-3) \times (-1) + 1 \times 0 = 15 \\ d_{12} &= 4 \times (-2) + (-3) \times 11 + 1 \times 6 = -35 \\ d_{13} &= 4 \times 0 + (-3) \times 0 + 1 \times 1 = 1 \\ D &= \begin{pmatrix} 15 & -35 & 1 \end{pmatrix} \end{aligned}$$

Now calculate $E = AB$.

$$\begin{aligned} E = AB &= \begin{pmatrix} 3 & -2 & 6 \\ -1 & 0 & 11 \end{pmatrix} \begin{pmatrix} 3 & -2 & 0 \\ -1 & 11 & 0 \\ 0 & 6 & 1 \end{pmatrix} \\ &= \begin{pmatrix} 9+2+0 & -6-22+36 & 0+0+6 \\ -3+0+0 & 2+0+66 & 0+0+11 \end{pmatrix} \\ &= \begin{pmatrix} 11 & 8 & 6 \\ -3 & 68 & 11 \end{pmatrix} \end{aligned}$$

Powers of matrices

When dealing with real numbers we calculate powers by repeated multiplication, e.g. $2^3 = 2 \times 2 \times 2 = 8$. The same principle applies to matrices, subject to the matrix being square; namely,

$$A^3 = A^2 A = AAA .$$

That A has to be square follows directly from the condition on matrix multiplication: '*... the number of columns in the first matrix must be equal to the number of rows in the second.*' Thus to calculate $A^2 = AA$ the number of columns of A must be equal to the number of rows of A, i.e. A must be square.

The identity matrix

The identity (or unit) matrix, I, is a diagonal matrix whose diagonal elements, a_{ii}, are all unity. The identity matrices of order 2×2 and 3×3 are

$$\begin{pmatrix} 1 & 0 \\ 0 & 1 \end{pmatrix} , \qquad \begin{pmatrix} 1 & 0 & 0 \\ 0 & 1 & 0 \\ 0 & 0 & 1 \end{pmatrix} .$$

The main feature of I is that for *any* square matrix A,

$$IA = AI = A .$$

This is one of the few cases where matrix multiplication is commutative. The identity matrix I is the matrix equivalent of unity in the real number system.

The determinant of a matrix

The determinant of a square matrix A is usually written as $|A|$ or $\det A$. To calculate the determinant of a matrix we replace the brackets with vertical lines and proceed exactly as in section 1.2 in the method of determinants. The determinant of a product of two matrices is the product of the individual determinants,

$$\det(AB) = \det A \det B .$$

EXERCISES

1.10. If $A = \begin{pmatrix} 1 & 2 \\ 0 & 1 \end{pmatrix}$ find A^2, A^3 and deduce A^n .

1.11. Carry out the matrix multiplication

$$\begin{pmatrix} 4 & -3 & 1 \\ 2 & 0 & 1 \\ 1 & 1 & -1 \end{pmatrix} \begin{pmatrix} 3 & -2 & 0 \\ -1 & 11 & 0 \\ 0 & 6 & 1 \end{pmatrix} .$$

1.12. If $A = \begin{pmatrix} 3 & 2 \\ 1 & 1 \end{pmatrix}$, $B = \begin{pmatrix} 1 & -2 \\ -1 & 3 \end{pmatrix}$ find det$A$, det$B$, AB and BA.

1.5.3 Inverse matrices

A square matrix A is said to be *invertible*, or to have an inverse, if there exists a matrix B such that

$$AB = BA = I .$$

Such a matrix is unique and the matrix B is called the inverse of A and is denoted by A^{-1}. Only square matrices may have an inverse. A square matrix has an inverse, or is non-singular, if and only if its determinant is non-zero, i.e. $|A| \neq 0$. The usefulness of the initial section on determinants should now become apparent.

The inverse of a matrix is derived entirely from determinants. It is necessary to define some more of the notation associated with matrix algebra.

The *minors* M_{ij} of a matrix are the *determinants* remaining when the ith row and j^{th} column are omitted. For example, a general 2×2 matrix is

$$\begin{pmatrix} a & b \\ c & d \end{pmatrix} .$$

There are 4 minors M_{11}, M_{12}, M_{21}, M_{22} associated with this matrix. If we omit a row and a column, then we are left with a single element, e.g. when we omit the first row and first column $M_{11} = d$ and similarly $M_{12} = c$, $M_{21} = b$, $M_2 = a$. The *matrix of co-factors* is defined as

$$\begin{pmatrix} M_{11} & -M_{12} \\ -M_{21} & M_{22} \end{pmatrix} .$$

Note the checkerboard pattern of + and − signs.

For a 3×3 matrix A, the minors will be defined as

$$M_{11} = \begin{vmatrix} a_{22} & a_{23} \\ a_{32} & a_{33} \end{vmatrix}, \quad M_{12} = \begin{vmatrix} a_{21} & a_{23} \\ a_{31} & a_{33} \end{vmatrix}, \quad \text{etc.}$$

and the matrix of co-factors is

$$\begin{pmatrix} M_{11} & -M_{12} & M_{13} \\ -M_{21} & M_{22} & -M_{23} \\ M_{31} & -M_{32} & M_{33} \end{pmatrix} .$$

Problem 1.8

Find the matrix of co–factors for the matrix $\begin{pmatrix} 3 & 2 & 1 \\ 1 & -1 & 4 \\ 0 & 6 & -2 \end{pmatrix}$.

Solution 1.8

The definition of the matrix of co–factors is

$$\begin{pmatrix} M_{11} & -M_{12} & M_{13} \\ -M_{21} & M_{22} & -M_{23} \\ M_{31} & -M_{32} & M_{33} \end{pmatrix}$$

$$= \begin{pmatrix} \begin{vmatrix} -1 & 4 \\ 6 & -2 \end{vmatrix} & -\begin{vmatrix} 1 & 4 \\ 0 & -2 \end{vmatrix} & -\begin{vmatrix} 1 & -1 \\ 0 & 6 \end{vmatrix} \\ -\begin{vmatrix} 2 & 1 \\ 6 & -2 \end{vmatrix} & \begin{vmatrix} 3 & 1 \\ 0 & -2 \end{vmatrix} & -\begin{vmatrix} 3 & 2 \\ 0 & 6 \end{vmatrix} \\ \begin{vmatrix} 2 & 1 \\ -1 & 4 \end{vmatrix} & -\begin{vmatrix} 3 & 1 \\ 1 & 4 \end{vmatrix} & -\begin{vmatrix} 3 & 2 \\ 1 & -1 \end{vmatrix} \end{pmatrix}$$

$$= \begin{pmatrix} -22 & 2 & 6 \\ 10 & -6 & -18 \\ 9 & -11 & -5 \end{pmatrix} .$$

Having found the matrix of co–factors, how does this help us find the inverse that we are seeking?

There is one more necessary definition. The *adjoint* of a matrix A, written adj$A$, is the transpose of the matrix of co–factors

$$\text{adj}A = \begin{pmatrix} M_{11} & -M_{12} & M_{13} \\ -M_{21} & M_{22} & -M_{23} \\ M_{31} & -M_{32} & M_{33} \end{pmatrix}^T = \begin{pmatrix} M_{11} & -M_{21} & M_{31} \\ -M_{12} & M_{22} & -M_{32} \\ M_{13} & -M_{23} & M_{33} \end{pmatrix} .$$

Finally, the inverse of a matrix A is defined by

$$A^{-1} = \frac{1}{\det A}\text{adj}A .$$

Returning to the general 2×2 matrix, the inverse is

$$A^{-1} = \frac{1}{ad - bc}\begin{pmatrix} d & -b \\ -c & a \end{pmatrix} .$$

This result can be easily verified by multiplication:

$$\begin{aligned} A^{-1}A &= \frac{1}{ad-bc}\begin{pmatrix} d & -b \\ -c & a \end{pmatrix}\begin{pmatrix} a & b \\ c & d \end{pmatrix} \\ &= \frac{1}{ad-bc}\begin{pmatrix} ad-bc & 0 \\ 0 & ad-bc \end{pmatrix} = \begin{pmatrix} 1 & 0 \\ 0 & 1 \end{pmatrix} = I\,. \end{aligned}$$

We have pre-multiplied A by A^{-1} and obtained the identity matrix. Post-multiplying, i.e. AA^{-1}, also gives I and hence we have found the inverse.

Summary of steps for finding the inverse of a matrix:

1. Calculate the determinant and check that it is non-zero.
2. Form the matrix of co–factors.
3. Calculate the adjoint matrix.
4. Calculate the inverse.

1.5.4 Matrix solution of linear equations

The inverse of a matrix can now be used to solve simultaneous linear equations as follows. Returning to the pair of equations in problem 1.1, we could write these in the form $AX = B$ where $A = \begin{pmatrix} 5 & -2 \\ 3 & 4 \end{pmatrix}$, $X = \begin{pmatrix} x \\ y \end{pmatrix}$, $B = \begin{pmatrix} 1 \\ 2 \end{pmatrix}$. Pre-multiplying this matrix equation by A^{-1} gives

$$\begin{aligned} A^{-1}AX &= A^{-1}B \\ IX &= A^{-1}B \\ X &= A^{-1}B \end{aligned}$$

and so the solution is obtained by multiplying B by the inverse of A.

Problem 1.9

Use matrices to solve the following pair of linear equations

$$\begin{aligned} 5x - 2y &= 1 \\ 3x + 4y &= 2\,. \end{aligned}$$

Solution 1.9

We have already defined the matrices by

$$A=\begin{pmatrix}5 & -2\\ 3 & 4\end{pmatrix},\quad X=\begin{pmatrix}x\\ y\end{pmatrix},\quad B=\begin{pmatrix}1\\ 2\end{pmatrix}.$$

The inverse of A is given by

$$A^{-1}=\frac{1}{26}\begin{pmatrix}4 & 2\\ -3 & 5\end{pmatrix}$$

and so the solution to the pair of equations is

$$\begin{aligned}X &= A^{-1}B\\ \begin{pmatrix}x\\ y\end{pmatrix} &= \frac{1}{26}\begin{pmatrix}4 & 2\\ -3 & 5\end{pmatrix}\begin{pmatrix}1\\ 2\end{pmatrix}=\frac{1}{26}\begin{pmatrix}8\\ 7\end{pmatrix}=\begin{pmatrix}\frac{8}{26}\\ \frac{7}{26}\end{pmatrix}.\end{aligned}$$

The solution is $x=\frac{8}{26}$ and $y=\frac{7}{26}$.

EXERCISES

1.13. Find the inverse of $B=\begin{pmatrix}-1 & 0 & 2\\ -10 & -1 & 0\\ 3 & 8 & 2\end{pmatrix}$.

1.14. Write down the inverse of $\begin{pmatrix}4 & -5\\ 1 & 9\end{pmatrix}$ and hence solve the simultaneous linear equations

$$\begin{aligned}4x - 5y &= -23\\ x + 9y &= 66\ .\end{aligned}$$

1.15. Use matrices to solve the example in the introduction to this chapter.

1.16. Use matrices to solve the simultaneous linear equations

$$\begin{aligned}5x + y - 2z &= -1\\ x + y + z &= -3\\ 7x + 8y &= -7\ .\end{aligned}$$

2 Vector Algebra

2.1 Introduction

Physical properties generally belong to one of two main groups, *scalars* or *vectors*. A vector is usually described as a quantity with *magnitude* and *direction*. For example, suppose a car travels northeast at 50 km/h. The *speed* of the car, 50 km/h, is a scalar quantity. The *velocity* of the car is a vector quantity, since it has a magnitude, 50 km/h, and a direction, northeast.

These notions of magnitude and direction allow us to represent vectors visually as directed line segments which can be combined by joining them 'nose to tail', as in figure 2.1. A vector joining two points A and B, fixed in space,

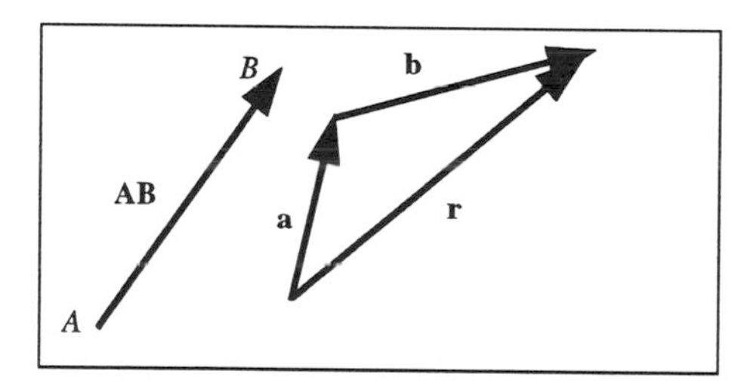

Figure 2.1. Vector notation

is a *position vector* denoted by $\mathbf{AB}$ where the order of the letters indicates the vector goes from A to B. Thus $\mathbf{BA}$ would indicate a vector going from B to A.

More general vectors are called *free vectors* and their position in space is irrelevant. They are usually denoted by a single lower-case letter in bold type, such as $\mathbf{r}$. In a diagram an arrow indicates the direction of the vector.
The *magnitude* of a position vector is denoted by $|\mathbf{AB}|$ or just AB and for a free vector the magnitude is denoted by $|\mathbf{r}|$ or r. The notation $\mathbf{r} = \mathbf{a} + \mathbf{b}$

indicates that $\mathbf{r}$, the *resultant* vector, is the sum of the vectors $\mathbf{a}$ and $\mathbf{b}$.
The product of a vector $\mathbf{r}$ and a scalar k is another vector $k\mathbf{r}$. Its magnitude is k times that of $\mathbf{r}$. If k is positive $k\mathbf{r}$ has the same direction as $\mathbf{r}$ and if k is negative its direction is the opposite of $\mathbf{r}$. In general, the equation $\mathbf{a} = k\mathbf{b}$ signifies that $\mathbf{a}$ and $\mathbf{b}$ are parallel vectors and the magnitude of $\mathbf{a}$ is $|k|$ times the magnitude of $\mathbf{b}$.

2.2 Algebraic representation

Geometric or graphical methods are limited by the accuracy of a scaled diagram. An analytical approach is often more convenient and more accurate. For applied problems the best approach is to use a combination of both methods – draw a sketch to visualise the problem and then use an analytical method to solve the problem accurately.

Analytic methods use a technique called *vector resolution*, where vectors are resolved into components. A vector can be resolved in many ways but the most convenient way is to use *rectangular* components aligned with the x, y and z coordinate axes.

Any point in space may be specified by reference to three mutually perpendicular axes Ox, Oy and Oz, as in figure 2.2, forming a righthanded system (rotate Ox through 90° to Oy and a right handed screw would move in the Oz direction). If P is a general point in space, with coordinates (x, y, z), then the

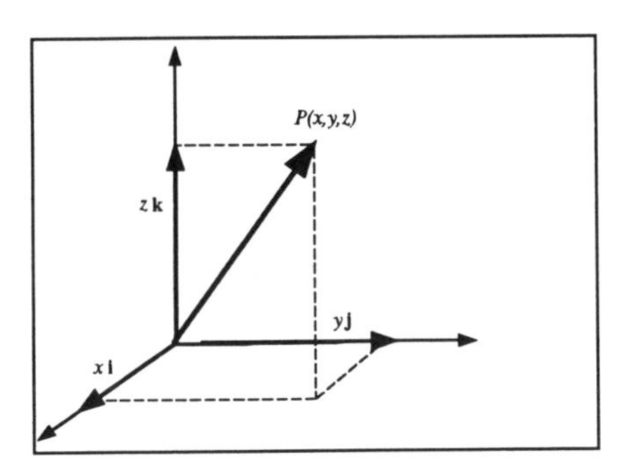

Fig. 2.2. Three-dimensional resolution

vector $\mathbf{OP}$ or $\mathbf{r}$ can be written as

$$\mathbf{OP} = \mathbf{r} = x\mathbf{i} + y\mathbf{j} + z\mathbf{k} \ ,$$

where $\mathbf{i}$, $\mathbf{j}$ and $\mathbf{k}$ are *unit vectors*, unit displacements, in the x, y and z coordinate directions respectively. The magnitude of $\mathbf{r}$ is $r = |\mathbf{r}| = \sqrt{x^2 + y^2 + z^2}$ and a unit vector in the same direction as $\mathbf{r}$ is

$$\hat{\mathbf{r}} = \frac{x\mathbf{i} + y\mathbf{j} + z\mathbf{k}}{\sqrt{x^2 + y^2 + z^2}} \,.$$

The use of ^ is standard notation to denote a unit vector.
Vectors are often written in matrix notation and referred to as column vectors

$$\mathbf{r} = \begin{pmatrix} x \\ y \\ z \end{pmatrix} .$$

However, to save space they are often written as the transpose of a row vector

$$\mathbf{r} = (\ x,\ \ y,\ \ z\ )^T .$$

Problem 2.1

If the points P, Q and R have coordinates (2,3,5), (3,1,2) and (2,-2,-1) express the vectors $\mathbf{OP}$, $\mathbf{OQ}$ and $\mathbf{PR}$ in terms of the unit vectors $\mathbf{i}$, $\mathbf{j}$ and $\mathbf{k}$ and determine a unit vector in the direction of $\mathbf{OP}$.

Solution 2.1

As unit displacements in the x, y and z directions are given by $\mathbf{i}$, $\mathbf{j}$ and $\mathbf{k}$ respectively, the vector linking two points is the product of the difference in coordinates and the appropriate unit vector.

$$\begin{aligned}
\mathbf{OP} &= (2-0)\mathbf{i} + (3-0)\mathbf{j} + (5-0)\mathbf{k} = 2\mathbf{i} + 3\mathbf{j} + 5\mathbf{k} \\
\mathbf{OQ} &= (3-0)\mathbf{i} + (1-0)\mathbf{j} + (2-0)\mathbf{k} = 3\mathbf{i} + \mathbf{j} + 2\mathbf{k} \\
\mathbf{PR} &= (2-2)\mathbf{i} + (-2-3)\mathbf{j} + (-1-5)\mathbf{k} = -5\mathbf{j} - 6\mathbf{k} \,.
\end{aligned}$$

The addition of vectors given in component form is obtained by adding the corresponding components. For example, if $\mathbf{a} = \mathbf{i} + 2\mathbf{j} - 3\mathbf{k}$ and $\mathbf{b} = 4\mathbf{i} - 2\mathbf{k}$ then the sum can be written in any of the following ways

$$\begin{aligned}
\mathbf{a} + \mathbf{b} &= (1+4)\mathbf{i} + (2+0)\mathbf{j} + (-3-2)\mathbf{k} \\
&= 5\mathbf{i} + 2\mathbf{j} - 5\mathbf{k} \\
\mathbf{a} + \mathbf{b} &= \begin{pmatrix} 1 \\ 2 \\ -3 \end{pmatrix} + \begin{pmatrix} 4 \\ 0 \\ -2 \end{pmatrix} = \begin{pmatrix} 5 \\ 2 \\ -5 \end{pmatrix} \\
\mathbf{a} + \mathbf{b} &= (\ 1,\ \ 2,\ \ -3\ )^T + (\ 4,\ \ 0,\ \ -2\ )^T \\
&= (\ 5,\ \ 2,\ \ -5\ )^T .
\end{aligned}$$

EXERCISES

2.1. Given that $\mathbf{a} = 3\mathbf{i} - \mathbf{j} - \mathbf{k}$, $\mathbf{b} = \mathbf{i} + \mathbf{j} + 5\mathbf{k}$ and $\mathbf{c} = -\mathbf{i} + 2\mathbf{k}$ find the vectors

$$\text{(a) } 2\mathbf{a} - \mathbf{b} \qquad \text{(b) } \mathbf{a} + \mathbf{b} + \mathbf{c} \qquad \text{(c) } 5\mathbf{a} - \mathbf{c} + 8\mathbf{b}\,.$$

2.2. Prove that the following are unit vectors

$$\text{(a) } \frac{1}{\sqrt{2}}(-\mathbf{j} + \mathbf{k}) \qquad \text{(b) } -\frac{3}{5}\mathbf{i} - \frac{4}{5}\mathbf{j}\,.$$

2.3. For $\mathbf{a}$, $\mathbf{b}$ and $\mathbf{c}$ given in exercise 2.1 above, determine $|\mathbf{a}|$, $|\mathbf{b}|$ and $|\mathbf{c}|$ and hence construct unit vectors $\hat{\mathbf{a}}$, $\hat{\mathbf{b}}$ and $\hat{\mathbf{c}}$.

2.3 Linear independence

We have already seen that if $\mathbf{a} = k\mathbf{b}$ then $\mathbf{a}$ and $\mathbf{b}$ are parallel vectors and we say that $\mathbf{a}$ is *linearly dependent* on $\mathbf{b}$. The set of points whose position vectors are linearly dependent on $\mathbf{b}$ is a line through the origin in the direction of $\mathbf{b}$.

We define a set $\{\mathbf{a}, \mathbf{b}\}$ of non-zero vectors to be linearly dependent if we can find scalars k and l, not both zero, such that $k\mathbf{a} + l\mathbf{b} = \mathbf{0}$. If the only solution to $k\mathbf{a} + l\mathbf{b} = \mathbf{0}$ is $k = l = 0$ then we say that the pair of vectors is linearly independent. The set of points whose position vectors are linearly dependent on $\{\mathbf{a}, \mathbf{b}\}$ is the plane through the origin containing the vectors $\mathbf{a}$ and $\mathbf{b}$. Every point on this plane has a position vector of the form $k\mathbf{a} + l\mathbf{b}$.

The extension to this is to say that the set $\{\mathbf{a}, \mathbf{b}, \mathbf{c}\}$ is linearly independent if the only way we can satisfy $k\mathbf{a} + l\mathbf{b} + m\mathbf{c} = \mathbf{0}$ is to have $k = l = m = 0$. In three dimensional space all vectors can be written as $k\mathbf{a} + l\mathbf{b} + m\mathbf{c}$ and a set of four vectors is always linearly dependent.

The set $\{\mathbf{a}, \mathbf{b}, \mathbf{c}\}$, when linearly independent, is called a *basis* because any vector $\mathbf{r}$ can be written

$$\mathbf{r} = x\mathbf{a} + y\mathbf{b} + z\mathbf{c}$$

and (x, y, z) are the components of $\mathbf{r}$ relative to the basis.

The ordinary cartesian components are those related to the basis $\{\mathbf{i}, \mathbf{j}, \mathbf{k}\}$ where the vectors are unit vectors along axes all at right angles to each other. Such a basis is called *ortho-normal.*

2.4 The scalar product

The *scalar product* is also referred to as the *inner product* or *dot product*. Consider vectors $\mathbf{a}$ and $\mathbf{b}$ where the angle between them is θ, as in figure 2.3.

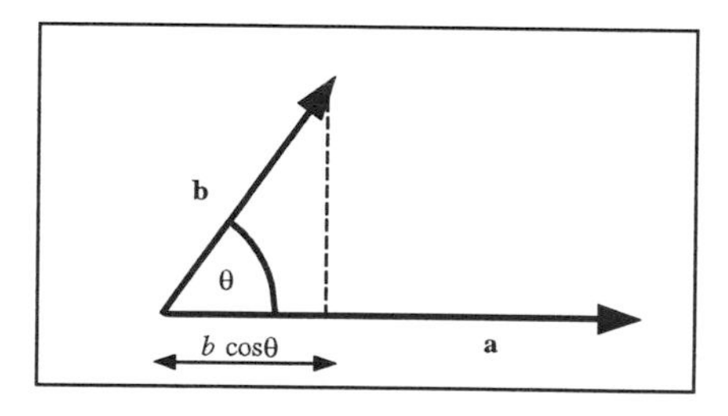

Fig. 2.3. The scalar projection of $\mathbf{b}$ onto $\mathbf{a}$ is $b\cos\theta$

The scalar product is defined as

$$\mathbf{a}.\mathbf{b} = ab\cos\theta$$

where a and b are the magnitudes of $\mathbf{a}$ and $\mathbf{b}$.
Given $\mathbf{a}$ and $\mathbf{b}$ in component form:

$$\mathbf{a} = a_1\mathbf{i} + a_2\mathbf{j} + a_3\mathbf{k}\ , \qquad \mathbf{b} = b_1\mathbf{i} + b_2\mathbf{j} + b_3\mathbf{k}$$

the scalar product is also defined by

$$\mathbf{a}.\mathbf{b} = a_1b_1 + a_2b_2 + a_3b_3\ .$$

What does the scalar product represent?

Suppose we replace $\mathbf{a}$ by a unit vector $\hat{\mathbf{n}}$. Then $\mathbf{b}.\hat{\mathbf{n}}$ is the component of $\mathbf{b}$ in the direction of $\hat{\mathbf{n}}$. We could define a unit vector in the direction of $\mathbf{a}$ by $\hat{\mathbf{a}} = \frac{1}{a}\mathbf{a}$ and then $\hat{\mathbf{a}}.\mathbf{b}$ represents the *scalar projection* of $\mathbf{b}$ onto $\mathbf{a}$. So the scalar product is the product of the magnitude of $\mathbf{a}$ and this projection of $\mathbf{b}$ onto $\mathbf{a}$. Note that if the vectors are perpendicular then there is zero scalar projection.

The scalar product obeys the following rules:

1. $\mathbf{a}.\mathbf{b} = \mathbf{b}.\mathbf{a}$
2. $k(\mathbf{a}.\mathbf{b}) = (k\mathbf{a}).\mathbf{b} = \mathbf{a}.(k\mathbf{b}) = k(\mathbf{b}.\mathbf{a})$
3. If $\mathbf{a}$ and $\mathbf{b}$ are perpendicular then $\cos\theta = 0$ and so $\mathbf{a}.\mathbf{b} = 0$
4. If $\mathbf{a}.\mathbf{b} = 0$ then either $\mathbf{a} = \mathbf{0}$ or $\mathbf{b} = \mathbf{0}$ or $\mathbf{a}$ and $\mathbf{b}$ are perpendicular
5. If $\mathbf{a} = a_1\mathbf{i} + a_2\mathbf{j} + a_3\mathbf{k}$ then $\mathbf{a}.\mathbf{a} = a_1^2 + a_2^2 + a_3^2 = a^2$
6. $\mathbf{i}.\mathbf{i} = \mathbf{j}.\mathbf{j} = \mathbf{k}.\mathbf{k} = 1$
7. $\mathbf{i}.\mathbf{j} = \mathbf{j}.\mathbf{k} = \mathbf{k}.\mathbf{i} = 0$

Problem 2.2

If $\mathbf{a} = \mathbf{i} + 3\mathbf{j} + 4\mathbf{k}$ and $\mathbf{b} = 2\mathbf{i} - \mathbf{j} + 6\mathbf{k}$ find the scalar product $\mathbf{a.b}$ and the acute angle (to 2 decimal places) between $\mathbf{a}$ and $\mathbf{b}$.

Solution 2.2

Using components

$$\mathbf{a.b} = 1 \times 2 + 3 \times (-1) + 4 \times 6 = 23\ .$$

$$\begin{aligned}
\text{Also} \qquad a &= \sqrt{1^2 + 3^2 + 4^2} = \sqrt{26} \\
b &= \sqrt{2^2 + (-1)^2 + 6^2} = \sqrt{41}\ . \\
\text{Therefore} \qquad \cos\theta &= \frac{\mathbf{a.b}}{ab} = \frac{23}{\sqrt{26 \times 41}} = 0.7044 \\
\theta &= 45.21^o \ \text{ or } \ 0.789\,\text{rad.}
\end{aligned}$$

EXERCISES

2.4. Find the scalar product $\mathbf{a.b}$ and the acute angle between the vectors when $\mathbf{a} = \mathbf{i} + \mathbf{j} + \mathbf{k}$ and $\mathbf{b} = \mathbf{i} - \mathbf{j}$.

2.5. Find the value of a such that the vectors $a\mathbf{i} - 2\mathbf{j} + \mathbf{k}$ and $2a\mathbf{i} + a\mathbf{j} - 4\mathbf{k}$ are perpendicular to each other.

2.6. Two sides of a triangle are represented by the vectors $\mathbf{a} = 2\mathbf{i} - \mathbf{j} + 3\mathbf{k}$ and $\mathbf{b} = \mathbf{i} + 2\mathbf{j} + 4\mathbf{k}$. Determine the three angles of the triangle and check that their sum is 180°.

2.7. The work done by a force $\mathbf{F}$ in displacing a body by the vector $\mathbf{d}$ is $\mathbf{F.d}$. Find the work done by a force of magnitude 5 newtons acting in the direction $(3\mathbf{i}, -\mathbf{j}, \mathbf{k})$ in moving a particle from the point $A(1, 0, 8)$ to the point $B(5, 3, 1)$ where distances are in metres.

2.5 The vector product

The *vector product* is also referred to as the *cross product.* Consider the two vectors $\mathbf{a}$ and $\mathbf{b}$ where the angle between them is θ, as shown in figure 2.4. The vector product is defined as

$$\mathbf{a} \times \mathbf{b} = ab \sin\theta \hat{\mathbf{n}}\ ,$$

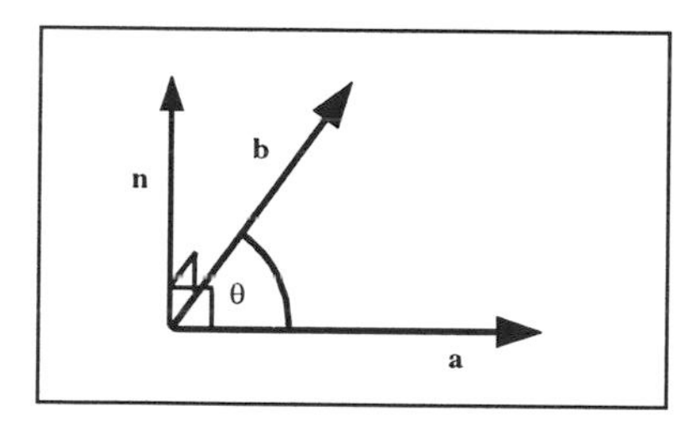

Fig. 2.4. The vector product $\mathbf{a} \times \mathbf{b}$ is perpendicular to $\mathbf{a}$ and $\mathbf{b}$

where $\hat{\mathbf{n}}$ is a unit vector perpendicular to both $\mathbf{a}$ and $\mathbf{b}$ such that $\mathbf{a}$, $\mathbf{b}$ and $\mathbf{a} \times \mathbf{b}$ form a right handed set. Notice that the result is a *vector*.

If $\mathbf{a} = a_1\mathbf{i} + a_2\mathbf{j} + a_3\mathbf{k}$ and $\mathbf{b} = b_1\mathbf{i} + b_2\mathbf{j} + b_3\mathbf{k}$ then the vector product is given by the third-order determinant (see chapter 1).

$$\mathbf{a} \times \mathbf{b} = \begin{vmatrix} \mathbf{i} & \mathbf{j} & \mathbf{k} \\ a_1 & a_2 & a_3 \\ b_1 & b_2 & b_3 \end{vmatrix}$$

$$= (a_2b_3 - a_3b_2)\mathbf{i} - (a_1b_3 - a_3b_1)\mathbf{j} + (a_1b_2 - a_2b_1)\mathbf{k} .$$

The following rules apply:

1. If $\mathbf{a} = \mathbf{0}$ or $\mathbf{b} = \mathbf{0}$ then $\mathbf{a} \times \mathbf{b} = 0$.
2. If $\mathbf{a}$ and $\mathbf{b}$ are parallel then $\mathbf{a} \times \mathbf{b} = 0$.
3. $\mathbf{a} \times \mathbf{b} = -(\mathbf{b} \times \mathbf{a})$.
4. $k(\mathbf{a} \times \mathbf{b}) = (k\mathbf{a}) \times \mathbf{b} = \mathbf{a} \times (k\mathbf{b})$.
5. $\mathbf{i} \times \mathbf{i} = \mathbf{j} \times \mathbf{j} = \mathbf{k} \times \mathbf{k} = 0$.
6. $\mathbf{i} \times \mathbf{j} = \mathbf{k}, \mathbf{j} \times \mathbf{k} = \mathbf{i}, \mathbf{k} \times \mathbf{i} = \mathbf{j}$.

Problem 2.3

If the vertices of a triangle are given by $A(0, 1, 1)$, $B(2, 0, 0)$ and $C(1, 3, 1)$ find $\mathbf{AB} \times \mathbf{BC}$.

Solution 2.3

First write the two vectors in component form

$$\mathbf{AB} = 2\mathbf{i} - \mathbf{j} - \mathbf{k} \qquad \mathbf{BC} = -\mathbf{i} + 3\mathbf{j} + \mathbf{k} .$$

Then evaluate the determinant

$$\mathbf{AB} \times \mathbf{BC} = \begin{vmatrix} \mathbf{i} & \mathbf{j} & \mathbf{k} \\ 2 & -1 & -1 \\ -1 & 3 & 1 \end{vmatrix} = 2\mathbf{i} - \mathbf{j} + 5\mathbf{k} .$$

EXERCISES

2.8. Find the vector product $\mathbf{a}\times\mathbf{b}$ when $\mathbf{a} = \mathbf{i}+\mathbf{j}-\mathbf{k}$ and $\mathbf{b} = 2\mathbf{i}-\mathbf{j}+3\mathbf{k}$.

2.9. If $A(0,1,1)$, $B(1,3,1)$ and $C(-1,-2,3)$ are the three vertices of a triangle, find a unit vector that is perpendicular to $\mathbf{AB}$ and $\mathbf{BD}$, where D is the midpoint of AC.

2.10. The moment $\mathbf{M}$ of a force $\mathbf{F}$ about a point P is given by $\mathbf{M} = \mathbf{r}\times\mathbf{F}$ where $\mathbf{r}$ is the vector from P to the point of application of the force. A force given by $\mathbf{F} = 3\mathbf{i}+2\mathbf{j}-4\mathbf{k}$ is applied at the point $P(1,-1,2)$. Find the moment of $\mathbf{F}$ about the point $(2,-1,3)$.

2.6 Triple products

There are various ways of forming products of three vectors $\mathbf{a}$, $\mathbf{b}$ and $\mathbf{c}$. The *triple scalar product* is the combination of a scalar product and a vector product

$$\mathbf{a}.(\mathbf{b}\times\mathbf{c}) = \mathbf{a}.(bc\sin\theta\hat{\mathbf{n}}) = abc\sin\theta\cos\phi$$

where θ is the angle between $\mathbf{b}$ and $\mathbf{c}$, and ϕ is the smaller angle between $\mathbf{a}$ and $\mathbf{b}\times\mathbf{c}$. Notice that the brackets are not strictly necessary because the alternative order $(\mathbf{a}.\mathbf{b})\times\mathbf{c}$ does not make sense because $(\mathbf{a}.\mathbf{b})$ is a scalar and so cannot be used in a vector product.

This result has the following geometric application. Consider the parallelepiped shown in figure 2.5 where the edges are represented by $\mathbf{a}$, $\mathbf{b}$ and $\mathbf{c}$.

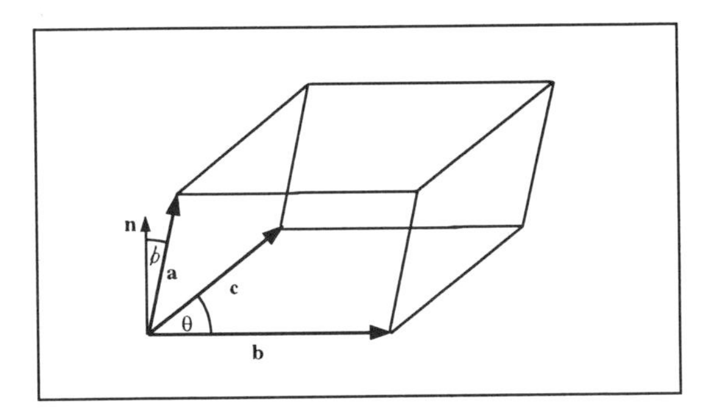

Fig. 2.5. Parallelepiped volume

The height of the parallelepiped is given by $a\cos\phi$ and the area of the base is given by $bc\sin\theta$. Using these relations we can see that $\mathbf{a}.(\mathbf{b}\times\mathbf{c})$ = volume of parallelepiped.

Trying to calculate the volume using the expression $abc \sin\theta \cos\phi$ is difficult if the angles θ and ϕ have to be calculated. If the vectors are given in terms of cartesian components it is more convenient to calculate a determinant.

If $\mathbf{a} = a_1\mathbf{i} + a_2\mathbf{j} + a_3\mathbf{k}$, $\mathbf{b} = b_1\mathbf{i} + b_2\mathbf{j} + b_3\mathbf{k}$, $\mathbf{c} = c_1\mathbf{i} + c_2\mathbf{j} + c_3\mathbf{k}$

then
$$\begin{aligned} \mathbf{b}\times\mathbf{c} &= \begin{vmatrix} \mathbf{i} & \mathbf{j} & \mathbf{k} \\ b_1 & b_2 & b_3 \\ c_1 & c_2 & c_3 \end{vmatrix} \\ &= \mathbf{i}(b_2c_3 - c_2b_3) - \mathbf{j}(b_1c_3 - c_1b_3) + \mathbf{k}(b_1c_2 - c_1b_2)\ . \end{aligned}$$

Now if we use the definition of the scalar product

$$\begin{aligned} \mathbf{a}.(\mathbf{b}\times\mathbf{c}) &= (a_1\mathbf{i} + a_2\mathbf{j} + a_3\mathbf{k}) \\ &\qquad .(\mathbf{i}(b_2c_3 - c_2b_3) - \mathbf{j}(b_1c_3 - c_1b_3) + \mathbf{k}(b_1c_2 - c_1b_2)) \\ &= a_1(b_2c_3 - c_2b_3) - a_2(b_1c_3 - c_1b_3) + a_3(b_1c_2 - c_1b_2) \\ &= \begin{vmatrix} a_1 & a_2 & a_3 \\ b_1 & b_2 & b_3 \\ c_1 & c_2 & c_3 \end{vmatrix}\ . \end{aligned}$$

As long as we keep the cyclic order of the vectors the result is the same

$$\begin{aligned} \mathbf{a}.(\mathbf{b}\times\mathbf{c}) &= \mathbf{c}.(\mathbf{a}\times\mathbf{b}) = \mathbf{b}.(\mathbf{c}\times\mathbf{a}) \\ \text{but}\quad \mathbf{a}.(\mathbf{b}\times\mathbf{c}) &= -\mathbf{a}.(\mathbf{c}\times\mathbf{b})\ . \end{aligned}$$

Problem 2.4

Find the volume of the parallelepiped whose adjacent edges are given by

$$\mathbf{a} = \mathbf{i} + 2\mathbf{j} - \mathbf{k}, \quad \mathbf{b} = -\mathbf{i} + \mathbf{j}, \quad \mathbf{c} = -\mathbf{j} + 2\mathbf{k}\ .$$

Solution 2.4

Substitute the components into the third order determinant.

$$\begin{aligned} \mathbf{a}.(\mathbf{b}\times\mathbf{c}) &= \begin{vmatrix} a_1 & a_2 & a_3 \\ b_1 & b_2 & b_3 \\ c_1 & c_2 & c_3 \end{vmatrix} = \begin{vmatrix} 1 & 2 & -1 \\ -1 & 1 & 0 \\ 0 & -1 & 2 \end{vmatrix} \\ &= 1(2-0) - 2(-2-0) - 1(1-0) = 5\ . \end{aligned}$$

Thus the volume is 5 cubic units.

The *triple vector product* of the vectors $\mathbf{a}$, $\mathbf{b}$ and $\mathbf{c}$ is denoted $\mathbf{a} \times (\mathbf{b} \times \mathbf{c})$.

If $\mathbf{a} = a_1\mathbf{i} + a_2\mathbf{j} + a_3\mathbf{k}, \ \mathbf{b} = b_1\mathbf{i} + b_2\mathbf{j} + b_3\mathbf{k}, \ \mathbf{c} = c_1\mathbf{i} + c_2\mathbf{j} + c_3\mathbf{k}$

$$\text{then} \qquad \mathbf{b} \times \mathbf{c} = \begin{vmatrix} \mathbf{i} & \mathbf{j} & \mathbf{k} \\ b_1 & b_2 & b_3 \\ c_1 & c_2 & c_3 \end{vmatrix}$$

$$= \mathbf{i}(b_2c_3 - c_2b_3) - \mathbf{j}(b_1c_3 - c_1b_3) + \mathbf{k}(b_1c_2 - c_1b_2) \ .$$

If we repeat the process we get

$$\mathbf{a} \times (\mathbf{b} \times \mathbf{c}) = \begin{vmatrix} \mathbf{i} & \mathbf{j} & \mathbf{k} \\ a_1 & a_2 & a_3 \\ b_2c_3 - c_2b_3 & -(b_1c_3 - c_1b_3) & b_1c_2 - c_1b_2 \end{vmatrix}$$

The proof of the following result is left as exercise 2.12.
The triple vector product can be expressed in the following form

$$\mathbf{a} \times (\mathbf{b} \times \mathbf{c}) = (\mathbf{a}.\mathbf{c})\mathbf{b} - (\mathbf{a}.\mathbf{b})\mathbf{c} \ .$$

Similarly,

$$(\mathbf{a} \times \mathbf{b}) \times \mathbf{c} = (\mathbf{a}.\mathbf{c})\mathbf{b} - (\mathbf{b}.\mathbf{c})\mathbf{a} \ .$$

Thus, except in the special case when $\mathbf{b}$ is perpendicular to both $\mathbf{a}$ and $\mathbf{c}$, and hence $\mathbf{a}.\mathbf{b} = \mathbf{b}.\mathbf{c} = 0$, we have

$$\mathbf{a} \times (\mathbf{b} \times \mathbf{c}) \neq (\mathbf{a} \times \mathbf{b}) \times \mathbf{c} \ .$$

Problem 2.5

If $\mathbf{a} = 2\mathbf{i} - \mathbf{j} + \mathbf{k}$, $\mathbf{b} = \mathbf{i} + 2\mathbf{j} - \mathbf{k}$ and $\mathbf{c} = \mathbf{j} + 2\mathbf{k}$ find the triple vector products (a) $\mathbf{a} \times (\mathbf{b} \times \mathbf{c})$ and (b) $(\mathbf{a} \times \mathbf{b}) \times \mathbf{c}$.

Solution 2.5

(a) To find $\mathbf{a} \times (\mathbf{b} \times \mathbf{c})$, we consider

$$\mathbf{b} \times \mathbf{c} = \begin{vmatrix} \mathbf{i} & \mathbf{j} & \mathbf{k} \\ 1 & 2 & -1 \\ 0 & 1 & 2 \end{vmatrix} = 5\mathbf{i} - 2\mathbf{j} + \mathbf{k}$$

and then we find the cross product of this vector with $\mathbf{a}$:

$$\mathbf{a} \times (\mathbf{b} \times \mathbf{c}) = \begin{vmatrix} \mathbf{i} & \mathbf{j} & \mathbf{k} \\ 2 & -1 & 1 \\ 5 & -2 & 1 \end{vmatrix} = \mathbf{i} + 3\mathbf{j} + \mathbf{k} \ .$$

(b) To find $(\mathbf{a} \times \mathbf{b}) \times \mathbf{c}$, we calculate

$$\mathbf{a} \times \mathbf{b} = \begin{vmatrix} \mathbf{i} & \mathbf{j} & \mathbf{k} \\ 2 & -1 & 1 \\ 1 & 2 & -1 \end{vmatrix} = -\mathbf{i} + 3\mathbf{j} + 5\mathbf{k} \, .$$

Now we find the cross product of this vector with $\mathbf{c}$:

$$(\mathbf{a} \times \mathbf{b}) \times \mathbf{c} = \begin{vmatrix} \mathbf{i} & \mathbf{j} & \mathbf{k} \\ -1 & 3 & 5 \\ 0 & 1 & 2 \end{vmatrix} = \mathbf{i} + 2\mathbf{j} - \mathbf{k} \, .$$

Note that $\mathbf{a} \times (\mathbf{b} \times \mathbf{c}) \neq (\mathbf{a} \times \mathbf{b}) \times \mathbf{c}$.

EXERCISES

2.11. Find the scalar triple product $\mathbf{a}.(\mathbf{b} \times \mathbf{c})$ when

$$\mathbf{a} = 2\mathbf{i} - 3\mathbf{j} + \mathbf{k} \, , \ \mathbf{b} = 3\mathbf{i} + \mathbf{j} + 2\mathbf{k} \, , \ \mathbf{c} = \mathbf{i} + 4\mathbf{j} - 2\mathbf{k} \, .$$

2.12. Prove that $\mathbf{a} \times (\mathbf{b} \times \mathbf{c}) = (\mathbf{a}.\mathbf{c})\mathbf{b} - (\mathbf{a}.\mathbf{b})\mathbf{c}$

2.7 Differentiation of vector functions

Let the curve in figure 2.6 be given by the parametric equations

$$x = f(t) \, , \qquad y = g(t) \, .$$

The vector $\mathbf{r} = x\mathbf{i} + y\mathbf{j} = f(t)\mathbf{i} + g(t)\mathbf{j}$ which joins the origin to the point $P(x, y)$ of the curve is called the position vector of P. If t denotes time then

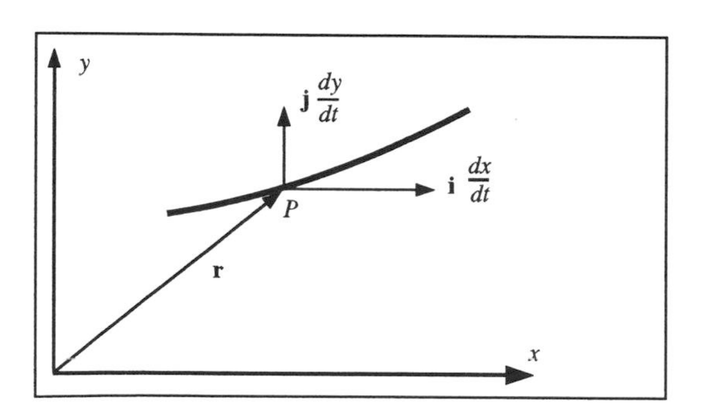

Fig. 2.6. Vector differentiation

the derivative of $\mathbf{r}$ with respect to t is given by

$$\frac{d\mathbf{r}}{dt} = \frac{dx}{dt}\mathbf{i} + \frac{dy}{dt}\mathbf{j}$$

and represents the velocity of a particle on the curve. Similarly, the second derivative with respect to t represents the acceleration of the particle at P. In three dimensions $\mathbf{r} = x\mathbf{i} + y\mathbf{j} + z\mathbf{k}$, with velocity and acceleration similarly defined.

Problem 2.6

A particle moves along the curve $x = 4\cos t$, $y = 4\sin t$ and $z = 6t$ where t represents time. Find the magnitude of the velocity and acceleration of the particle at the times $t = 0$ and $t = \frac{\pi}{2}$.

Solution 2.6

Let $P(x, y, z)$ be a point on the curve and let $\mathbf{r} = x\mathbf{i} + y\mathbf{j} + z\mathbf{k}$ be its position vector. In terms of t, $\mathbf{r}$ can be written

$$\mathbf{r} = 4\cos t\mathbf{i} + 4\sin t\mathbf{j} + 6t\mathbf{k}\ .$$

$$\begin{aligned} \text{Then} \quad \mathbf{v} &= \frac{d\mathbf{r}}{dt} = -4\sin t\mathbf{i} + 4\cos t\mathbf{j} + 6\mathbf{k} \\ \mathbf{a} &= \frac{d^2\mathbf{r}}{dt^2} = -4\cos t\mathbf{i} - 4\sin t\mathbf{j}\ . \end{aligned}$$

At $t = 0$: $\mathbf{v} = 4\mathbf{j} + 6\mathbf{k}$, $|\mathbf{v}| = 2\sqrt{13}$, $\mathbf{a} = -4\mathbf{i}$, $|\mathbf{a}| = 4$.
At $t = \frac{\pi}{2}$: $\mathbf{v} = -4\mathbf{i} + 6\mathbf{k}$, $|\mathbf{v}| = 2\sqrt{13}$, $\mathbf{a} = -4\mathbf{j}$, $|\mathbf{a}| = 4$.
The particle's motion describes a circle of radius 4 units in the x, y plane with a constant velocity in the z direction, rather like a coiled spring.

EXERCISES

2.13. A particle has position vector $\mathbf{r} = t^2\mathbf{i} - \sin(4t)\mathbf{j} + e^{2t}\mathbf{k}$. Determine the velocity and acceleration at $t = 0$.

2.14. A vector $\mathbf{n}$ is given by $\mathbf{n} = \cos\theta\mathbf{i} + \sin\theta\mathbf{j}$ where θ depends on t.
  (a) verify that $\mathbf{n}$ is a unit vector
  (b) determine $\frac{d}{dt}(\mathbf{n})$
  (c) verify that $\frac{d}{dt}(\mathbf{n})$ and $\mathbf{n}$ are perpendicular.

2.15. For $\mathbf{r} = t^2\mathbf{i} - t\mathbf{j} + (2t+1)\mathbf{k}$ and $\mathbf{s} = (2t-3)\mathbf{i} + \mathbf{j} - t\mathbf{k}$ determine

$$\text{(a)}\ \frac{d}{dt}(\mathbf{r}.\mathbf{s})\ ,\ \text{(b)}\ \frac{d}{dt}(\mathbf{r}\times\mathbf{s})\ ,\ \text{(c)}\ \frac{d}{dt}(|\mathbf{r}+\mathbf{s}|)\ ,\ \text{(d)}\ \frac{d}{dt}(\mathbf{r}\times\frac{d}{dt}\mathbf{s})\ .$$

3
Complex Numbers

3.1 Introduction

Complex numbers are an extension of the real number system, just as real numbers are an extension of the set of rational numbers. Complex numbers make it possible to solve any quadratic equation. For example, if we attempt to find the roots of the quadratic $x^2 - 2x + 5 = 0$ using the standard formula we get

$$x = \frac{2 \pm \sqrt{4-20}}{2} = 1 \pm \frac{\sqrt{-16}}{2} = 1 \pm 2\sqrt{-1}\,.$$

If we *define* the *imaginary unit* by the symbol i, where $\mathrm{i} = \sqrt{-1}$ and hence $\mathrm{i}^2 = -1$, then the solutions to the above quadratic are the *complex numbers* $1 + 2\mathrm{i}$ and $1 - 2\mathrm{i}$.

We now consider a general complex number z as having the form $z = a + b\mathrm{i}$ where a and b are *real* numbers. The real part of z, also written Re z, is a and Im $z = b$ is the imaginary part of z.

The most common areas of application for complex numbers are in electrical circuit theory and mechanical vector analysis. In electrical circuit theory i is the accepted symbol for electric current and so in engineering mathematics $\mathrm{j} = \sqrt{-1}$ is used instead.

Rather than thinking of complex numbers as an entirely new concept it is helpful to note the similarities with vectors and matrices. A complex number can be thought of as being a vector with two components; this idea is reinforced when we represent complex numbers graphically.

3.2 Algebra of complex numbers

If $z_1 = a_1 + b_1\text{i}$ and $z_2 = a_2 + b_2\text{i}$ are two complex numbers they satisfy the following rules:

1. $z_1 + z_2 = (a_1 + a_2) + (b_1 + b_2)\text{i}$
2. $z_1 - z_2 = (a_1 - a_2) + (b_1 - b_2)\text{i}$
3. $kz_1 = ka_1 + kb_1\text{i}$, where k is a real number
4. If $z_1 = z_2$ then $a_1 = a_2$ and $b_1 = b_2$
5. $z_1 z_2 = (a_1 a_2 - b_1 b_2) + (a_1 b_2 + a_2 b_1)\text{i}$
6. $\bar{z}_1 = a_1 - b_1\text{i}$ is the complex *conjugate* of z_1, sometimes written as z_1^*
7. $z_1 \bar{z}_1 = a_1^2 + b_1^2$ is a real number. Note $z\bar{z} \geq 0$
8. The modulus of z_1 is denoted $|z_1|$ and is defined by $|z_1|^2 = z_1 \bar{z}_1$
9. Addition is associative and commutative:
$$z_1 + (z_2 + z_3) = (z_1 + z_2) + z_3 \ , \qquad z_1 + z_2 = z_2 + z_1$$
10. Multiplication is associative and commutative:
$$z_1(z_2 z_3) = (z_1 z_2) z_3 \ , \qquad z_1 z_2 = z_2 z_1 \ .$$

Division of a complex number by a real number is straightforward; for example,

$$\frac{6 - 4\text{i}}{2} = 3 - 2\text{i} \ , \qquad \frac{a + b\text{i}}{a} = 1 + \frac{b}{a}\text{i} \ .$$

To divide one complex number by another, we multiply both numerator and denominator by the complex conjugate of the denominator. This converts the denominator into a real number and the division can then be carried out as above.

$$\frac{7 - 4\text{i}}{4 + 3\text{i}} = \frac{7 - 4\text{i}}{4 + 3\text{i}} \times \frac{4 - 3\text{i}}{4 - 3\text{i}} = \frac{(7 - 4\text{i})(4 - 3\text{i})}{4^2 + 3^2} = \frac{16 - 37\text{i}}{25} = \frac{16}{25} - \frac{37}{25}\text{i} \ .$$

Problem 3.1

Find a and b if $(a + b) + (a - b)\text{i} = 8 + 2\text{i}$ and hence obtain the modulus of the complex number $z = a + b\text{i}$.

Solution 3.1

Equating real and imaginary parts we have

$$a + b = 8 \ , \qquad a - b = 2 \ ,$$

which has the solution $a = 5$ and $b = 3$; hence

$$|z| = \sqrt{a^2 + b^2} = \sqrt{34} \ .$$

EXERCISES

3.1. Express (a) $\frac{4-5i}{1+2i}$ and (b) $\frac{1+6i}{7i}$ in the form $a+bi$.
3.2. Find a and b if $(a+b)+(a-b)i=(2+5i)^2+i(2-3i)$.
3.3. If $z=\frac{2+i}{1-i}$ find the real and imaginary parts of $z+\frac{1}{z}$.
3.4. Solve the following equation for z

$$z^2+\bar{z}^2+2z=4+4i\ .$$

3.3 Graphical representation

Complex numbers are represented on an *Argand diagram* and are considered as ordered pairs of real numbers, $x+yi\equiv(x,y)$. This notation allows us to treat complex numbers in the same way as vectors written in component form. The real part is plotted along the horizontal axis and the imaginary part on the vertical axis.

With this in mind the complex number $3+2i$ is represented by OA in figure 3.1. Note the similarity to plotting the position vector $3\mathbf{i}+2\mathbf{j}$.

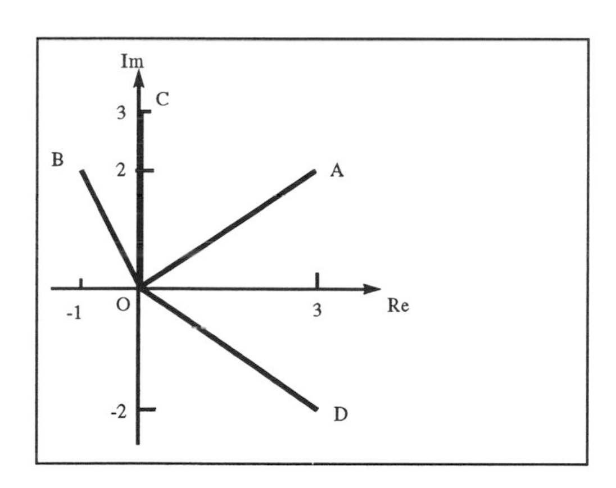

Fig. 3.1. Graphical representation of complex numbers

Figure 3.1 also shows:

1. OB represents the complex number $-1+2i$
2. OC represents the complex number $0+3i$
3. OD represents the complex number $3-2i$

Notice that OD is the complex conjugate of OA and is shown by reflection in the horizontal axis.

To add two complex numbers graphically we follow the same procedure used for vectors, either the parallelogram rule or the completion of the triangle when two complex numbers are drawn nose to tail. Figure 3.2 illustrates the addition of the complex numbers $z_1 = 1 + 3\mathrm{i}$ and $z_2 = 4 + \mathrm{i}$.

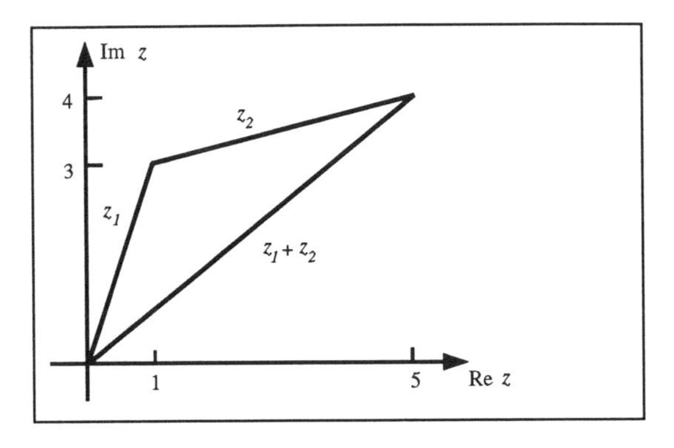

Fig. 3.2. Addition of complex numbers

3.4 Polar form

Let $z = x + y\mathrm{i}$ be the general complex number shown on the Argand diagram in figure 3.3. If θ is the angle the complex number makes with the real axis and

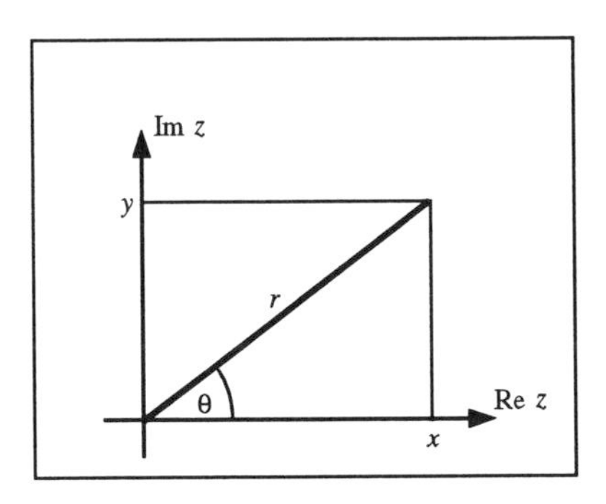

Fig. 3.3. Polar form of complex numbers

r is the length of the position vector $\mathbf{r}$ then we have

$$x = r\cos\theta\ , \qquad y = r\sin\theta\ .$$

From the right angled triangle we have $r^2 = x^2 + y^2$ and $\tan\theta = \dfrac{y}{x}$ and so

$$z = r(\cos\theta + \mathrm{i}\sin\theta) \ .$$

Thus we have represented the complex number z by its magnitude and the angle it makes with the real axis. This is called the *polar form* of the complex number z where r is the *modulus* and θ is called the *argument*, abbreviated to arg. This is usually written as $r\angle\theta$.

For any complex number $z \neq 0$ there is only one value of θ in $0 \leq \theta < 2\pi$. However, any interval of length 2π, including the standard choice $-\pi \leq \theta < \pi$, can be used. Any particular choice, decided upon in advance, is called the *principal range*, and the value of θ in this range is called the *principal value* or *principal argument*. The argument θ can be expressed in radians or degrees provided the units are made clear.

Problem 3.2

Express the complex number $4 - 3\mathrm{i}$ in polar form.

Solution 3.2

Initially it may be helpful to draw a diagram.

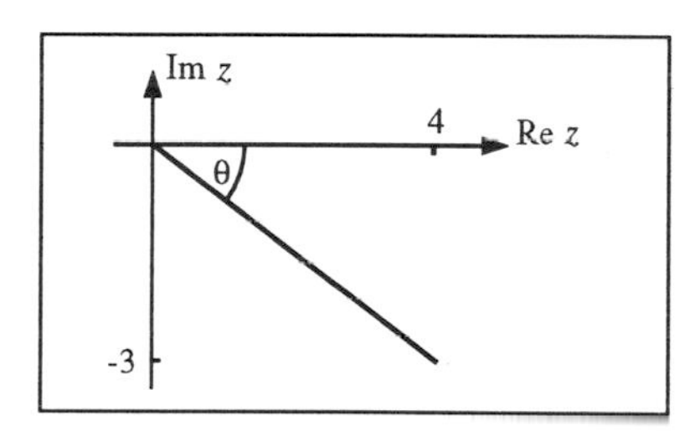

Fig. 3.4. Polar form of $4 - 3\mathrm{i}$

The modulus is $r = \sqrt{4^2 + (-3)^2} = 5$ and the argument is calculated using the inverse tangent

$$\theta = \tan^{-1}\left(\frac{-3}{4}\right) = -36.87^\circ \ .$$

Notice that this angle is in the range $-180^\circ \leq \theta < 180^\circ$. If we had declared our principal range to be $0 \leq \theta < 360^\circ$ then $\theta = 360 - 36.87 = 323.13^\circ$ would be the principal value.

Problem 3.3

Express the complex number $4\angle 65^\circ$ in the form $a + b\mathrm{i}$.

Solution 3.3

$$\begin{aligned} 4\angle 65^\circ &= 4(\cos 65^\circ + \mathrm{i}\sin 65^\circ) \\ &= 4(0.4226 + 0.9063\mathrm{i}) \\ &= 1.6904 + 3.6252\mathrm{i}\,. \end{aligned}$$

Multiplication and division of complex numbers in polar form are defined by

$$\begin{aligned} z_1 z_2 &= (r_1\angle\theta_1)(r_2\angle\theta_2) \\ &= (r_1 r_2)\angle(\theta_1+\theta_2) \\ \frac{z_1}{z_2} &= \frac{r_1\angle\theta_1}{r_2\angle\theta_2} \\ &= \frac{r_1}{r_2}\angle(\theta_1-\theta_2)\,. \end{aligned}$$

The proofs of these results are exercises.

Problem 3.4

Calculate $z_1 z_2$ and $\frac{z_1}{z_2}$ when $z_1 = 3\angle 30^\circ$ and $z_2 = 2\angle 47^\circ$, if the principal range is $0 \le \theta < 360^\circ$.

Solution 3.4

$$\begin{aligned} z_1 z_2 &= (r_1 r_2)\angle(\theta_1+\theta_2) = 6\angle 77^\circ\,, \\ \frac{z_1}{z_2} &= \frac{r_1}{r_2}\angle(\theta_1-\theta_2) = \frac{3}{2}\angle -17^\circ = \frac{3}{2}\angle 343^\circ\,. \end{aligned}$$

3.5 Exponential form

Many mathematical functions can be expressed as a series with positive integer powers of x (see the section in chapter 6 on Taylor Series). Examples include

$$\begin{aligned} \mathrm{e}^x &= 1 + x + \frac{x^2}{2!} + \frac{x^3}{3!} + \frac{x^4}{4!} + \ldots \\ \sin x &= x - \frac{x^3}{3!} + \frac{x^5}{5!} - \frac{x^7}{7!} + \ldots \\ \cos x &= 1 - \frac{x^2}{2!} + \frac{x^4}{4!} - \frac{x^6}{6!} + \ldots \end{aligned}$$

For the sine and cosine expansions x is measured in radians.

Suppose for the moment that we accept the above series are still valid when x is replaced by $\mathrm{i}\theta$, where θ is measured in radians. Then we have

$$\begin{aligned} \mathrm{e}^{\mathrm{i}\theta} &= 1+(\mathrm{i}\theta)+\frac{(\mathrm{i}\theta)^2}{2!}+\frac{(\mathrm{i}\theta)^3}{3!}+\frac{(\mathrm{i}\theta)^4}{4!}+\ldots \\ &= 1+\mathrm{i}\theta-\frac{\theta^2}{2!}-\frac{\mathrm{i}\theta^3}{3!}+\frac{\theta^4}{4!}+\ldots \\ &= \left(1-\frac{\theta^2}{2!}+\frac{\theta^4}{4!}+\ldots\right)+\mathrm{i}\left(\theta-\frac{\theta^3}{3!}+\frac{\theta^5}{5!}\ldots\right) \\ &= \cos\theta+\mathrm{i}\sin\theta \\ \text{Therefore} \quad r\mathrm{e}^{\mathrm{i}\theta} &= r(\cos\theta+\mathrm{i}\sin\theta)\ . \end{aligned}$$

This is known as the exponential form of a complex number.

Euler's formula is simply a statement of the exponential form for complex numbers. It states that by *assuming* that the infinite series expansion of e^x holds when $x=\mathrm{i}\theta$ radians we can write

$$\mathrm{e}^{\mathrm{i}\theta}=\cos\theta+\mathrm{i}\sin\theta\ .$$

In general, if $z=x+\mathrm{i}y$ then we define

$$\mathrm{e}^z=\mathrm{e}^{x+\mathrm{i}y}=\mathrm{e}^x\mathrm{e}^{\mathrm{i}y}=\mathrm{e}^x(\cos y+\mathrm{i}\sin y)$$

For the particular case when $y=0$, z is real, this reduces to e^x as expected.

To multiply two complex numbers in exponential form we proceed as we would for real indices.

$$\left(r_1\mathrm{e}^{\mathrm{i}\theta_1}\right)\left(r_2\mathrm{e}^{\mathrm{i}\theta_2}\right)=r_1r_2\mathrm{e}^{(\mathrm{i}\theta_1+\mathrm{i}\theta_2)}=r_1r_2\mathrm{e}^{\mathrm{i}(\theta_1+\theta_2)}$$

Similarly, division of two complex numbers in exponential form is given by

$$\frac{r_1\mathrm{e}^{\mathrm{i}\theta_1}}{r_2\mathrm{e}^{\mathrm{i}\theta_2}}=\frac{r_1}{r_2}\mathrm{e}^{\mathrm{i}\theta_1-\mathrm{i}\theta_2}=\frac{r_1}{r_2}\mathrm{e}^{\mathrm{i}(\theta_1-\theta_2)}$$

provided $r_2\neq 0$.

Problem 3.5

Calculate the product z_1z_2 and express this product in the form $a+b\mathrm{i}$ if

$$z_1=3\mathrm{e}^{\mathrm{i}\pi/5} \qquad z_2=4\mathrm{e}^{\mathrm{i}\pi/10}\ .$$

Calculate a and b accurate to 3 decimal places.

Solution 3.5

$$\begin{aligned} z_1z_2 &= \left(3e^{i\pi/5}\right)\left(4e^{i\pi/10}\right) \\ &= 12e^{i\pi/5+i\pi/10} = 12e^{i3\pi/10} \\ &= 12\left(\cos\left(\frac{3\pi}{10}\right) + i\sin\left(\frac{3\pi}{10}\right)\right) = 7.053 + 9.708i \end{aligned}$$

EXERCISES

3.5. Convert $1 - \sqrt{3}i$ to polar form.
3.6. Multiply $5\angle 30°$ by $3\angle 15°$ and express the result in the form $a + bi$.
3.7. Divide $5\angle 320°$ by $2\angle 35°$ and express the result in the form $a + bi$.
3.8. Divide $3e^{i4\pi/5}$ by $3e^{i\pi/2}$ and express the result in the form $a + bi$.
3.9. Prove that if z_1 and z_2 are in polar form then $z_1z_2 = (r_1\angle\theta_1)(r_2\angle\theta_2)$.

3.6 De Moivre's theorem

If two complex numbers z_1 and z_2 are written in the form

$$\begin{aligned} z_1 &= r_1(\cos\theta_1 + i\sin\theta_1) \\ z_2 &= r_2(\cos\theta_2 + i\sin\theta_2) \end{aligned}$$

then we can write the product as

$$z_1z_2 = r_1r_2\left(\cos(\theta_1+\theta_2) + i\sin(\theta_1+\theta_2)\right) .$$

This result generalises for the product of n complex numbers

$$z_1z_2\ldots z_n = r_1r_2\ldots r_n\left(\cos(\theta_1+\theta_2+\ldots+\theta_n) + i\sin(\theta_1+\theta_2+\ldots+\theta_n)\right) .$$

If all the complex numbers are equal, $z_1 = z_2 = \ldots = z_n = r(\cos\theta + i\sin\theta)$ then this result simplifies to

$$z^n = r^n\left(\cos(n\theta) + i\sin(n\theta)\right) .$$

If z is written in exponential form De Moivre's theorem is

$$z^n = \left(re^{i\theta}\right)^n = r^ne^{in\theta} .$$

Problem 3.6

Show that $\cos\theta = \frac{1}{2}\left(e^{i\theta} + e^{-i\theta}\right)$ and $\sin\theta = \frac{1}{2i}\left(e^{i\theta} - e^{-i\theta}\right)$.

Solution 3.6

We have

$$e^{i\theta} = \cos\theta + i\sin\theta, \qquad e^{-i\theta} = \cos\theta - i\sin\theta .$$

Adding and subtracting respectively, we get

$$e^{i\theta} + e^{-i\theta} = 2\cos\theta, \qquad e^{i\theta} - e^{-i\theta} = 2i\sin\theta .$$

and these lead to the required results

$$\cos\theta = \frac{1}{2}\left(e^{i\theta} + e^{-i\theta}\right), \qquad \sin\theta = \frac{1}{2i}\left(e^{i\theta} - e^{-i\theta}\right) .$$

Problem 3.7

Prove that $\sin^3\theta = \frac{3}{4}\sin\theta - \frac{1}{4}\sin(3\theta)$.

Solution 3.7

We have

$$\begin{aligned}
\sin\theta &= \frac{1}{2i}\left(e^{i\theta} - e^{-i\theta}\right) \\
\sin^3\theta &= \frac{1}{8i^3}\left(e^{i\theta} - e^{-i\theta}\right)^3 \\
&= -\frac{1}{8i}\left(\left(e^{i\theta}\right)^3 - 3\left(e^{i\theta}\right)^2\left(e^{-i\theta}\right) + 3\left(e^{i\theta}\right)\left(e^{-i\theta}\right)^2 - \left(e^{-i\theta}\right)^3\right) \\
&= -\frac{1}{8i}\left(e^{3i\theta} - 3e^{i\theta} + 3e^{-i\theta} - e^{-3i\theta}\right) \\
&= \frac{3}{8i}\left(e^{i\theta} - e^{-i\theta}\right) - \frac{1}{8i}\left(e^{3i\theta} - e^{-3i\theta}\right) = \frac{3}{4}\sin\theta - \frac{1}{4}\sin(3\theta) .
\end{aligned}$$

3.6.1 Roots of unity

A complex number z such that $z^n = 1$ is called an nth *root of unity.*

If $z = r(\cos\theta + i\sin\theta)$ then $z^n = r^n(\cos n\theta + i\sin n\theta)$ follows from De Moivre's theorem. Now, $z^n = 1$ if and only if $r^n = 1$ *and* $n\theta = 2k\pi$ where k is an integer.

Hence the nth root of unity (of which there are n) has modulus 1 and its argument is a multiple of $\frac{2\pi}{n}$. On an Argand diagram these all lie on the circle of radius 1, centre the origin. Figure 3.5 shows the fifth roots of unity. The real root $z = 1$ is on the real axis and the others are equally spaced at $\frac{2\pi}{n}$ intervals; in this example $n = 5$ and the intervals are $\frac{2\pi}{5}$.

If $w = \cos(\frac{2\pi}{n}) + i\sin(\frac{2\pi}{n})$ then the n roots are given by $1, w^2, w^3, \ldots, w^{n-1}$.

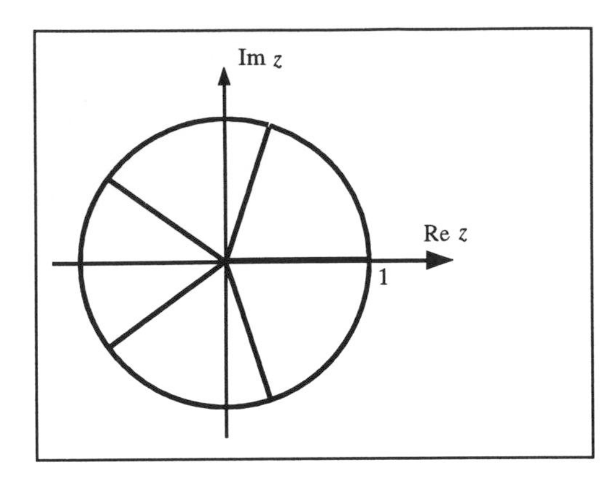

Fig. 3.5. Fifth roots of unity

3.6.2 Roots of complex numbers

We can generalise the roots of unity to the roots of an arbitrary complex number. A complex number w is called an nth root of another complex number z if $w^n = z$, or $w = z^{1/n}$.

If n is a positive integer then De Moivre's theorem gives

$$\begin{aligned} z^{1/n} &= (r(\cos(\theta) + \mathrm{i}\sin(\theta)))^{1/n} \\ &= r^{1/n}\left(\cos\frac{\theta + 2k\pi}{n} + \mathrm{i}\sin\frac{\theta + 2k\pi}{n}\right) . \end{aligned}$$

From this it follows that there are n different values for $z^{1/n}$, i.e. n different roots of z provided $z \neq 0$. Note that the arguments are displaced by θ and the modulus of the roots is $r^{1/n}$.

EXERCISES

3.10. Prove that $\cos^4\theta = \frac{1}{8}\cos(4\theta) + \frac{1}{2}\cos(2\theta) + \frac{3}{8}$.

3.11. Find the square roots of $-15 - 8\mathrm{i}$.

3.12. Find the third roots of $-1 + \mathrm{i}$.

4 Review of Differentiation Techniques

4.1 Introduction

Rollercoasters may differ in size and terror level but they are all based on the same principles. A car is pulled steadily up to the highest point and from then on gravity is the driving force. The car accelerates downwards before turning and twisting up and down smaller loops until the brakes are applied to bring the car to a standstill.

Before construction of a rollercoaster begins it is the subject of a mathematical simulation using models of the physical phenomena involved. The simulation is used to answer questions such as:

- How fast can the car go without undue risk of a major component failure?
- Will the forces generated dangerously affect the health of the passengers?
- Does the ride *appear* dangerous enough to generate customers?

Displaying parameters graphically gives an overall impression of the general behaviour and allows *estimates* of maximum speed and force, regions of maximum acceleration / deceleration (*rates of change* of speed), etc. but these estimates are subjective. Ideally we want a mathematical function that can be investigated analytically.

The calculus of one independent real variable uses the notation

$$y = f(x)$$

where y, the *dependent* variable, is a function of x, the *independent* variable.

Finding the *rate* at which y changes with respect to x is a typical problem of differential calculus. This chapter reviews the basic techniques which allow us to determine rates of change analytically. Chapter 6 then applies these techniques to problems of maxima/minima and optimisation.

Before we review the differentiation techniques there are a few preliminaries regarding functions.

Domain and range of functions

If $f(x)$ is a function of a real variable x, the *domain* of the function is the set of all allowable values of x. The set of all possible values of $f(x)$ is then called the *range* of f. For example, the domain of the function

$$f(x) = \frac{\sqrt{x^2 - 16}}{x^2}$$

are those real values such that we take the square root of a positive quantity, i.e. $x^2 - 16 \geq 0$ and the denominator is non-zero, i.e. $x^2 \neq 0$. Thus the domain is $|x| \geq 4$. The range of $f(x)$ is the set of positive real numbers, including zero.

Continuous and differentiable functions

The concept of a *limit* is a major difference between calculus and algebra. The value of the function $f(x)$ tends to a limiting value l as x tends to x_0, i.e.

$$\lim_{x \to x_0} f(x) = l$$

if all values of $f(x)$ for which x is close to x_0 differ negligibly little from l.

A function $f(x)$ is said to be *continuous at a point* x_0 if the following three conditions are satisfied:

$$f(x_0) \text{ is defined,} \qquad \lim_{x \to x_0} f(x) \text{ exists,} \qquad \lim_{x \to x_0} f(x) = f(x_0)\ .$$

The function is continuous on an interval if it is continuous at every point in the interval.

The derivative of a function $f(x)$ at a point x_0 is defined by

$$f'(x_0) = \lim_{h \to 0} \frac{f(x_0 + h) - f(x_0)}{h}$$

provided the limit exists.

The function is said to be *differentiable* over an interval if the derivative exists at every point in the interval.

The derivative of y with respect to x can appear in a variety of notations:

$$\frac{dy}{dx} \qquad \frac{d}{dx}(f(x)) \qquad f'(x) \qquad f' \qquad y'(x) \qquad y'\ .$$

Properties of derivatives

If k, l are constants and u, v are functions of x:

$$\frac{d}{dx}(ku \pm lv) = k\frac{du}{dx} \pm l\frac{dv}{dx}\ .$$

4.2 Differentiation of standard functions

Table 4.1 lists some common standard derivatives (a and n are constants).

As this chapter is a review of differentiation techniques we will begin by illustrating some of the results in table 4.1 using straightforward examples.

	$y = f(x)$	$\frac{dy}{dx} = f'(x)$		$y = f(x)$	$\frac{dy}{dx} = f'(x)$
D1	x^n	nx^{n-1}	D6	$\sinh(ax)$	$a\cosh(ax)$
D2	$\ln\lvert ax\rvert$	$\frac{1}{x}$	D7	$\cosh(ax)$	$a\sinh(ax)$
D3	e^{ax}	ae^{ax}	D8	$\tan^{-1} x$	$\frac{1}{x^2+1}$
D4	$\sin(ax)$	$a\cos(ax)$	D9	$\sin^{-1} x$	$\frac{1}{1-x^2}$
D5	$\cos(ax)$	$-a\sin(ax)$	D10	$\cosh^{-1} x$	$\frac{1}{\sqrt{x^2-1}}$

Table 4.1. Table of standard derivatives

Problem 4.1

Differentiate the following functions using table 4.1 and the basic properties of derivatives. Evaluate the derivative at the value of the independent variable indicated, to 3 decimal place accuracy in (c) and (d).

(a) $y = 2x^7 + \frac{7}{x^2} + 3$, $x = 1$

(b) $y = \sin(3x) - 4\cos(5x)$, $x = \pi/2$

(c) $z = e^{3t} - 2\ln 2t$, $t = 1$

(d) $u = 9\tan^{-1} t$, $t = 6$

Solution 4.1

(a) We differentiate each term separately using D1 in table 4.1. Note that $\frac{d}{dx}(c) = 0$ where c is a constant, corresponds to $n = 0$ in D1.

$$\begin{aligned} y &= 2x^7 + \frac{7}{x^2} + 3 = 2x^7 + 7x^{-2} + 3 \\ \frac{dy}{dx} &= \frac{d}{dx}(2x^7) + \frac{d}{dx}(7x^{-2}) + \frac{d}{dx}(3) \\ &= 14x^6 - 14x^{-3} + 0 = 14x^6 - \frac{14}{x^3} \end{aligned}$$

When $x = 1$ $\frac{dy}{dx} = 14 - 14 = 0$.

(b) Differentiation of sine and cosine are given by D4, D5.

$$\begin{aligned} y &= \sin(3x) - 4\cos(5x) \\ \frac{dy}{dx} &= 3\cos(3x) - 4(-5)\sin(5x) = 3\cos(3x) + 20\sin(5x) \end{aligned}$$

When $x = \frac{\pi}{2}$ $\frac{dy}{dx} = 3\cos(3\pi/2) + 20\sin(5\pi/2) = 20$.

(c) Differentiation of logs and exponents are given by D2, D3.

$$z = \mathrm{e}^{3t} - 2\ln 2t \qquad \frac{dz}{dt} = 3\mathrm{e}^{3t} - \frac{2}{t}$$

When $t = 1$ $\frac{dz}{dt} = 3\mathrm{e}^3 - 2 = 58.257$.

(d) Differentiation of inverse tangent is given by D8.

$$u = 9\tan^{-1} t \qquad \frac{du}{dt} = \frac{9}{t^2 + 1}$$

When $t = 6$ $\frac{du}{dt} = \frac{9}{36+1} = 0.243$.

EXERCISES

4.1. Differentiate the following functions and simplify your answers using square root signs or positive exponents only.
(a) $v = z + \frac{1}{z}$ (b) $y = x^{0.8} + \frac{1}{x^{0.2}}$ (c) $z = 2\sqrt{y} - \frac{4}{\sqrt{y}}$

4.2. Differentiate the following functions.
(a) $y = 3\cos(5x) - 3\mathrm{e}^{2x} + 6$ (c) $P = 2 - 24t + 3\sinh(4t)$
(b) $y = 6\ln(2x) - 2x^{-2}$ (d) $Q = \sin^2 x + \cos^2 x$

4.3. A car is travelling at a constant speed when the brakes are suddenly applied. The subsequent distance, s, travelled by the car is

$$s = 40t - 4t^2$$

where the time t is measured from the instant the brakes are applied. Find the speed of the car $v = \frac{ds}{dt}$. How long before the car stops?

4.4. The atmospheric pressure P pascals at a height h metres above sea-level is given by

$$P = P_0 e^{-kh}$$

where the constant $k = 1.25 \times 10^{-4}$ and $P_0 = 1.01 \times 10^5$ pascals is the pressure at sea-level. Find the rate of change of pressure with height at 3000 m above sea-level.

4.5. The angular displacement θ (radians) of a rotating disc is given by

$$\theta = 2\sin(3t)$$

where t is time (seconds). Find the angular velocity, $\frac{d\theta}{dt}$. What is the first value of t for which the velocity is zero?

4.3 Function of a function

Although table 4.1 is useful it is limited. For example, we expect the derivative of $y = \sin(7x + 0.01)$ to be very similar to the derivative of $y = \sin(7x)$ but table 4.1 is not general enough to indicate the difference.

The idea behind the function of a function rule (or *chain rule*) is to split a more complicated differentiation into simpler steps.

For example, suppose $y = (3x+1)^7$ and we define $u = 3x + 1$. Then $y = u^7$ and we now have two functions which are easy to differentiate

$$\frac{dy}{du} = 7u^6 \quad \text{and} \quad \frac{du}{dx} = 3\,.$$

The derivative we want is given by the product of these two derivatives.

$$\frac{dy}{dx} = \frac{dy}{du}\frac{du}{dx}\,.$$

Notice that the du's 'cancel' on the right hand side to leave $\frac{dy}{dx}$.

Problem 4.2

Find the derivatives of the following functions

(a) $y = 2\cos(\pi - 5x)$, (b) $y = \ln(1 + x^2)$, (c) $s = (2t^2 + 7)^{10}$.

Solution 4.2

For each function the substitution is 'u = bracketed term'.
(a) Let $u = \pi - 5x$; then $y = 2\cos u$,

$$\frac{du}{dx} = -5\,, \qquad \frac{dy}{du} = -2\sin u\,,$$
$$\frac{dy}{dx} = \frac{dy}{du}\frac{du}{dx} = (-2\sin u)(-5) = 10\sin u = 10\sin(\pi - 5x)\,.$$

(b) Let $u = 1 + x^2$; then $y = \ln u$,

$$\frac{du}{dx} = 2x\,, \qquad \frac{dy}{du} = \frac{1}{u}\,,$$
$$\frac{dy}{dx} = \frac{dy}{du}\frac{du}{dx} = \left(\frac{1}{u}\right)(2x) = \frac{2x}{u} = \frac{2x}{1+x^2}\,.$$

(c) Let $u = 2t^2 + 7$; then $s = u^{10}$,

$$\frac{du}{dt} = 4t\,, \qquad \frac{ds}{du} = 10u^9\,,$$
$$\frac{ds}{dt} = \frac{ds}{du}\frac{du}{dt} = (10u^9)(4t) = 40tu^9 = 40t(2t^2+7)^9\,.$$

EXERCISES

Differentiate the following functions.

4.6. $y = \sin(3x - \pi)$

4.7. $y = \ln(5 - 7x)$

4.8. $y = \mathrm{e}^{3x^2+1}$

4.9. $P = \cos^7 t$

4.10. The height $y(t)$ of a bungee jumper can be expressed in the form

$$y(t) = y_0 \sin\left(\sqrt{\frac{k}{m}}t - \frac{\pi}{2}\right)$$

where k represents the stiffness of the cord, m is the mass of the jumper and y_0 is the starting position of the jumper relative to the equilibrium position. Find the vertical speed of the jumper, $v = \frac{dy}{dt}$, at the times $t = 0$ seconds and $t = 3$ seconds given the following values: $y_0 = 20$ m, $m = 50$ kg, $k = 20$.

4.4 The product rule

So far we have differentiated standard functions or functions which can be made standard with an appropriate substitution. This section describes how to differentiate a *product* of two or more functions.

If u and v are both differentiable functions of x and if $y = uv$ then the product rule for differentiation is defined as

$$\frac{dy}{dx} = \frac{d}{dx}(uv) = v\frac{du}{dx} + u\frac{dv}{dx} .$$

To obtain the right-hand side we differentiate each function in turn, leaving the other alone. The product rule generalises easily to products of more than two functions. If w is also a differentiable function of x, and $y = uvw$ then

$$\frac{dy}{dx} = \frac{d}{dx}(uvw) = vw\frac{du}{dx} + uw\frac{dv}{dx} + uv\frac{dw}{dx} .$$

Notice that the order of the terms on the right hand side is immaterial.

When using the product rule it is best to follow a standard procedure.

1. *State* the two (or more) functions and then write down their derivatives.
2. Substitute the functions and derivatives *directly* into the product rule.
3. Simplify the result (if possible).

Problem 4.3

Differentiate $y = x^2 e^{2x}$, simplifying as far as possible.

Solution 4.3

This is a product of the two standard functions x^2 and e^{2x} and so the product rule is the appropriate method to use.

$$u = x^2 , \qquad v = e^{2x} ,$$

$$\frac{du}{dx} = 2x , \qquad \frac{dv}{dx} = 2e^{2x} .$$

Substitute these expressions into the product rule and simplify.

$$\begin{aligned}\frac{dy}{dx} &= v\frac{du}{dx} + u\frac{dv}{dx} = e^{2x}(2x) + x^2(2e^{2x}) \\ &= 2xe^{2x} + 2x^2e^{2x} = 2x(1+x)e^{2x} .\end{aligned}$$

Problem 4.4

Find the derivative of $y = x^2 \ln(1 + x^2)$ with respect to x.

Solution 4.4

Here y is obviously a product of the functions x^2 and $\ln(1+x^2)$ and so we use the product rule to find the derivative. For the derivative of $\ln(1+x^2)$ we use the result of problem 4.2(b).

$$u = x^2 \qquad v = \ln(1+x^2)$$

$$\frac{du}{dx} = 2x \qquad \frac{dv}{dx} = \frac{2x}{1+x^2}$$

Substitute these expressions into the product rule and simplify.

$$\begin{aligned}\frac{dy}{dx} &= v\frac{du}{dx} + u\frac{dv}{dx} \\ &= (2x)\ln(1+x^2) + x^2\left(\frac{2x}{1+x^2}\right) \\ &= 2x\ln(1+x^2) + \frac{2x^3}{1+x^2}\end{aligned}$$

EXERCISES

Use the product rule to find $\frac{dy}{dx}$ when

4.11. $y = x^3\cos(5x+2)$

4.12. $y = (x+6)^3 e^{5x+2}$

4.13. $y = e^x \sin(2x-1)$

4.14. $y = x^2 \ln x$

4.15. The power output P of a car engine is given by

$$P = Cne^{-nk}$$

where n is the engine speed in rpm and C and k are constants. Find the rate of change of power with respect to engine speed.

4.16. The current I in an electrical circuit varies with time t according to

$$I = e^{-3t}(0.4\cos(50t) + 0.1\sin(50t))$$

Calculate the value of $\frac{dI}{dt}$ when $t = 0.2$ seconds.

4.5 The quotient rule

If u and v are differentiable functions of x then the quotient rule is given by

$$\frac{d}{dx}\left(\frac{u}{v}\right) = \frac{v\frac{du}{dx} - u\frac{dv}{dx}}{v^2} \, .$$

This appears more complex than the product rule but uses the same four functions. Indeed, the numerator is the product rule with a minus sign instead of a plus sign which makes the order of the terms important.

Problem 4.5

Use the quotient rule to find $\frac{dy}{dx}$ where y is given by

$$y = \frac{e^{7x}}{x^2} \, .$$

(Note: in the introduction to this chapter we made some general statements about the assumed domains and ranges of functions. It is particularly important when we have a quotient that we exclude values of x which are zeros of the denominator. Here the function y is always positive but we must exclude $x = 0$. The domain of y is the set of real numbers except $x = 0$ and the range is the set of positive real numbers.)

Solution 4.5

Set the solution out in the same manner as the product rule. Write down the functions u and v and then their derivatives:

$$u = e^{7x} \, , \qquad v = x^2 \, ,$$

$$\frac{du}{dx} = 7e^{7x} \, , \qquad \frac{dv}{dx} = 2x \, .$$

Now substitute directly into the quotient rule and simplify.

$$\begin{aligned} \frac{dy}{dx} &= \frac{v\frac{du}{dx} - u\frac{dv}{dx}}{v^2} = \frac{x^2(7e^{7x}) - e^{7x}(2x)}{(x^2)^2} \\ &= \frac{(7x-2)xe^{7x}}{x^4} = \frac{(7x-2)e^{7x}}{x^3} \, . \end{aligned}$$

Problem 4.6

Use the quotient rule to find $\frac{dP}{dt}$ where P is given by

$$P = \frac{\cos t - 1}{\sin t + 3} \, .$$

(Note: the denominator satisfies $2 \leq \sin t + 3 \leq 4 \quad \forall t \in \Re$, and so we do not have to exclude any values of t from the domain of P on mathematical grounds.)

Solution 4.6

Write down the functions u and v and their derivatives:

$$u = \cos t - 1\,, \qquad v = \sin t + 3\,,$$

$$\frac{du}{dt} = -\sin t\,, \qquad \frac{dv}{dt} = \cos t\,.$$

Substitute these directly into the quotient rule and simplify.

$$\frac{dP}{dt} = \frac{v\frac{du}{dt} - u\frac{dv}{dt}}{v^2} = \frac{(\sin t + 3)(-\sin t) - (\cos t - 1)(\cos t)}{(\sin t + 3)^2}$$

$$= \frac{\cos t - 3\sin t - (\sin^2 t + \cos^2 t)}{(\sin t + 3)^2} = \frac{\cos t - 3\sin t - 1}{(\sin t + 3)^2}$$

EXERCISES

4.17. Use the quotient rule to find $\frac{dy}{dx}$ when

$$y = \frac{\ln(5 + x)}{5 + x}$$

and calculate the value of the derivative when $x = -4$.

4.18. Find the rate of change of P with respect to t if

$$P = \frac{e^{2t}}{1 + t^2}\,.$$

4.19. When the current through a transformer is x amps the efficiency of the transformer is given by

$$\epsilon = \frac{200x}{0.64x^2 + 200x + 350}\,.$$

Find the value of x such that $\frac{d\epsilon}{dx} = 0$.

4.20. The power P developed in a resistor R by a battery with emf E and internal resistance r is

$$P = \frac{e^2 R}{(R + r)^2}\,.$$

Find the rate of change of power with respect to R.

4.21. By writing $y = \frac{u}{v}$ as $y = uv^{-1}$ use the *product rule* to derive the quotient rule.

4.6 Higher derivatives

So far we have reviewed methods for finding the derivative $\frac{dy}{dx}$ of a function $y = f(x)$. More formally, this is the *first derivative*; but $\frac{dy}{dx}$ is itself a function of x and can usually be differentiated again.

If the first derivative is differentiable then its derivative is called the *second derivative* of y and usually denoted by one of the following

$$\frac{d^2y}{dx^2}, \quad f''(x), \quad y''(x), \quad y'' .$$

The process can be continued to find a *third derivative*,

$$\frac{d^3y}{dx^3}, \quad f'''(x), \quad y'''(x), \quad y''',$$

and so on. For higher derivatives it is easy to make mistakes with the 'dashed' notation and so the $\frac{d^ny}{dx^n}$ notation is recommended.

Note: the nth derivative at a point only exists if the original function and *all* derivatives up to the nth derivative are differentiable at the point.

Problem 4.7

Find the first three derivatives of the following functions and state any points for which a derivative does not exist.

$$\text{(a)}\ y = x^2 - 6x + 1, \qquad \text{(b)}\ y = \frac{2}{3}x^{3/2} .$$

Solution 4.7

(a)

$$y = x^2 - 6x + 1,$$

$$\frac{dy}{dx} = 2x - 6\,, \quad \frac{d^2y}{dx^2} = 2\,, \quad \frac{d^3y}{dx^3} = 0\,.$$

Notice that for an nth degree polynomial the derivatives of order $(n+1)$ and higher are zero.

(b)

$$y = \frac{2}{3}x^{\frac{3}{2}},$$

$$\frac{dy}{dx} = x^{\frac{1}{2}}\,, \quad \frac{d^2y}{dx^2} = \frac{1}{2x^{\frac{1}{2}}}\,, \quad \frac{d^3y}{dx^3} = -\frac{1}{4}x^{-\frac{3}{2}} = \frac{-1}{4x^{\frac{3}{2}}}\,.$$

The second derivative does not exist at $x = 0$ (division by zero) and so no higher derivative can exist at $x = 0$.

We shall see in chapter 6, when studying problems of maxima and minima, that the first and second derivatives are used most often.

EXERCISES

4.22. Find the first three derivatives of $y = \sin(2x) - \cos(3x)$.

4.23. Find the value of the second derivative of $y = e^t \sin t$ at $t = 0$.

4.7 Implicit differentiation

In every example so far, the dependent variable y has been expressed *explicitly* in terms of the independent variable x, i.e. in the form $y = f(x)$. Some functions are given in the *implicit* form $f(x, y) = 0$, for example

$$x^2 - xy + y = 0 .$$

This particular example can be rewritten in the form

$$y = \frac{x^2}{x-1}$$

and then differentiated using the quotient rule. An alternative is to use *implicit differentiation*. Remembering that y is a function of x we can differentiate as follows. Note the use of the product rule for the xy term.

$$\begin{aligned}
\frac{d}{dx}\left(x^2 - xy + y\right) &= \frac{d}{dx}(0) = 0, \\
\frac{d}{dx}(x^2) - \frac{d}{dx}(xy) + \frac{d}{dx}(y) &= 0, \\
2x - \left(y\frac{d}{dx}(x) + x\frac{d}{dx}(y)\right) + \frac{d}{dx}(y) &= 0, \\
2x - y - x\frac{dy}{dx} + \frac{dy}{dx} &= 0, \\
2x - y + (1-x)\frac{dy}{dx} &= 0, \\
\frac{dy}{dx} &= \frac{y-2x}{1-x} .
\end{aligned}$$

It is usually advisable to leave the derivative as an implicit function of y and x because of the difficulty in expressing y explicitly in terms of x.

Problem 4.8

Find the first derivative of y with respect to x when

$$y^2 - y = \sin x \, .$$

Solution 4.8

Differentiating implicitly

$$\begin{aligned} \frac{d}{dx}\left(y^2 - y - \sin x\right) &= \frac{d}{dx}(0) = 0, \\ \frac{d}{dx}(y^2) - \frac{d}{dx}(y) - \frac{d}{dx}(\sin x) &= 0, \\ 2y\frac{dy}{dx} - \frac{dy}{dx} - \cos x &= 0, \\ (2y-1)\frac{dy}{dx} - \cos x &= 0 \Rightarrow \frac{dy}{dx} = \frac{\cos x}{2y-1} \, . \end{aligned}$$

EXERCISES

Find the first derivatives of the following functions.

4.24. $x^3 - xy^2 = 0$

4.25. $\sin(xy) - x = 6$

4.8 Logarithmic differentiation

When the function is the product of several factors, some of which may be raised to a power, logarithmic differentiation is a useful extension of implicit differentiation and is illustrated in the following example.

Problem 4.9

Find the value (to 3 significant figures) of the first derivative of the following function when $x = 1$:

$$y = \frac{(x+1)^2(x+3)^{1/2}}{(x+6)^4} \quad , \quad x > 0 \, .$$

Solution 4.9

One possibility is to use a combination of the product rule and the quotient rule, but this will be long and tedious. An alternative, which is especially useful when a specific value is required, is to take logs of the equation and then to differentiate implicitly.

$$\begin{aligned}
\ln y &= \ln\left(\frac{(x+1)^2(x+3)^{1/2}}{(x+6)^4}\right), \\
\ln y &= 2\ln(x+1) + \frac{1}{2}\ln(x+3) - 4\ln(x+6) .
\end{aligned}$$

Now differentiating gives

$$\begin{aligned}
\frac{1}{y}\frac{dy}{dx} &= \frac{2}{x+1} + \frac{1/2}{x+3} - \frac{4}{x+6}, \\
\frac{dy}{dx} &= y\left(\frac{2}{x+1} + \frac{1/2}{x+3} - \frac{4}{x+6}\right), \\
\frac{dy}{dx} &= \frac{(x+1)^2(x+3)^{1/2}}{(x+6)^4}\left(\frac{2}{x+1} + \frac{1/2}{x+3} - \frac{4}{x+6}\right) .
\end{aligned}$$

We could remove the brackets by multiplying but as we only require the particular value when $x = 1$ we substitute at this stage.

$$\text{When } x = 1 \quad \frac{dy}{dx} = \frac{2^2 4^{1/2}}{7^4}\left(\frac{2}{2} + \frac{1/2}{4} - \frac{4}{7}\right) = 1.84 \times 10^{-3} .$$

EXERCISES

4.26. Differentiate the following function and find the value of $f'(2)$

$$f(x) = \frac{(x+7)^{3/2}(x-2)}{(2x-3)^8} .$$

4.27. Differentiate with respect to x and simplify

$$f(x) = (2x+3)^x .$$

5 Review of Integration Techniques

5.1 Introduction

Integration is usually viewed as the reverse of differentiation. Although most functions can be differentiated, considerably fewer functions can be integrated.

An integral is composed of an *integral sign*, the *integrand* – the function to be integrated – and the *variable* the integration is with respect to. We write

$$\int f(x)\, dx$$

where the integrand is $f(x)$ and the 'dx' indicates the integration is with respect to x, in the same way that $\frac{d}{dx}$ implies differentiation with respect to x.

When we differentiate a function we find a rate of change to which additive constants do not contribute. The result is that many functions have the same derivative. For example

$$\frac{d}{dx}(x^2) = \frac{d}{dx}(x^2+5) = \frac{d}{dx}(x^2-3) = 2x\ .$$

When we try to reverse the process using integration we don't know *precisely* which function we started with (we don't know the value of the constant).

The integral of the function $f(x)$ is another function $F(x)$ whose derivative is $f(x)$. $F(x)$ is called the *indefinite* integral (or anti–derivative) of $f(x)$, i.e.

$$F(x) = \int f(x)\, dx \qquad \text{such that} \qquad F'(x) = f(x)\ .$$

We have already seen that the indefinite integral of $f(x) = 2x$ is

$$F(x) = \int 2x\, dx = x^2 + c$$

where c is an arbitrary constant called the *constant of integration*; the indefinite integral is unique up to an additive constant.

The definite integral

The integral notation evolved from a summation process for the calculation of the area under a curve, the integral sign being an elongated S in 'Sum'.

The notion of area can be useful when trying to grasp the concept of definite integrals and the role of the limits. Suppose the interval $a \leq x \leq b$ is split into

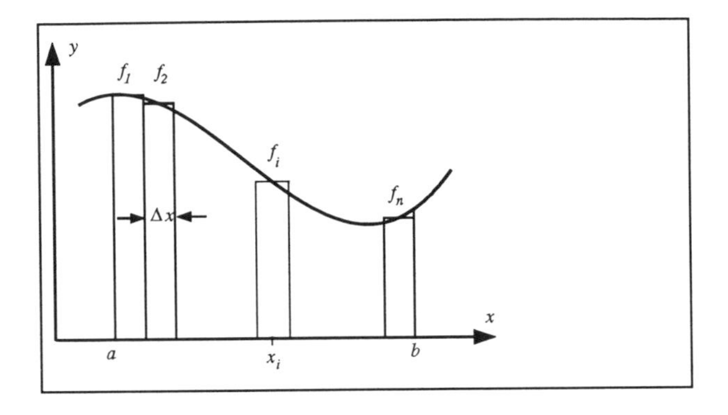

Fig. 5.1. Area under a curve

n equal elements of width Δx and a series of rectangles is generated whose height $f_i = f(x_i)$ is the value of $f(x)$ at the midpoint x_i of the interval, see figure 5.1. The area of each rectangle is $f(x_i) \times \Delta x \quad (1 \leq i \leq n)$ and the sum, S, of the areas of these rectangles *approximates* the area between $y = f(x)$ and the x-axis

$$S = \sum_{i=1}^{i=n} f(x_i)\Delta x \, .$$

This approximation improves as $n \to \infty$, $\Delta x \to 0$, and the limiting value is defined to be the *definite* integral of the function over the interval a to b

$$\int_a^b f(x)\, dx = \lim_{\Delta x \to 0} \sum_{i=1}^{i=n} f(x_i)\Delta x \, .$$

The fundamental theorem of integral calculus

If $f(x)$ is continuous over the interval $a \leq x \leq b$ and $F(x)$ is any indefinite integral of $f(x)$ then

$$\int_a^b f(x)\, dx = [F(x)]_a^b = F(b) - F(a)$$

where the notation $[F(x)]_a^b$ means evaluate $F(x)$ at the upper limit, $x = b$, then subtract the value of $F(x)$ at the lower limit, $x = a$. Notice that we omitted the constant of integration. Including the constant has no effect:

$$\int_a^b f(x)\,dx = [F(x)+c]_a^b = (F(b)+c) - (F(a)+c) = F(b) - F(a)\,.$$

We have already seen that the derivative of x^2 is $2x$ so, for example,

$$\int_1^3 2x\,dx = \left[x^2\right]_1^3 = 9 - 1 = 8\,.$$

$F(a)$ represents the enclosed area up to the point $x = a$. Similarly, $F(b)$ represents the area up to the point $x = b$. Therefore $F(b) - F(a)$ is the area bounded by the curve and the x–axis over the interval $a \le x \le b$.

If the upper and lower limits in a definite integral are identical, the integral is zero – the area under the curve vanishes as the points come together. From the fundamental theorem of integral calculus we have

$$\int_a^a f(x)\,dx = [F(x)]_a^a = F(a) - F(a) = 0\,.$$

Properties of integrals

If k, l are constants and $f(x)$, $g(x)$ are functions of x:

$$\int k\,f(x) \pm l\,g(x)\,dx = k\int f(x)\,dx \pm l\int g(x)\,dx\,.$$

Continuity of integrals

If an interval is split into two (or more) subintervals, then the integral over the whole interval is equal to the sum of the values over the subintervals

$$\int_a^b f(x)\,dx + \int_b^c f(x)\,dx = \int_a^c f(x)\,dx, \qquad a < b < c\,.$$

This result is logical when thought of as the sum of two neighbouring areas. Swapping the limits changes the sign of the integral

$$\int_b^a f(x)\,dx = F(a) - F(b) = -(F(b) - F(a)) = -\int_a^b f(x)\,dx\,.$$

To visualise this as an area – if we move from right to left along the x-axis we are reducing, or subtracting, an area.

We must be extra careful with definite integrals and ensure that the integrand is defined at all points between the limits. Consider the integrals:

$$\int_1^3 \frac{1}{x}\,dx, \qquad \int_{-1}^3 \frac{1}{x}\,dx\,.$$

There is no problem with the first integral but the interval for the second includes the point $x = 0$, at which point the integrand is not defined.

5.2 Integration of standard functions

Table 5.1 lists some of the more common standard indefinite integrals. In each case the constant of integration must be added.

	$f(x)$	$\int f(x)\,dx$	
I1	x^n	$\frac{x^{n+1}}{n+1}$,	$n \neq -1$
I2	$\frac{1}{x}$	$\ln\lvert x\rvert$	
I3	e^{ax}	$\frac{1}{a}\, e^{ax}$	
I4	$\sin(ax)$	$-\frac{1}{a}\cos(ax)$	
I5	$\cos(ax)$	$\frac{1}{a}\sin(ax)$	
I6	$\tan(ax)$	$-\frac{1}{a}\ln\lvert\cos(ax)\rvert$	
I7	$\sinh(ax)$	$\frac{1}{a}\cosh(ax)$	
I8	$\cosh(ax)$	$\frac{1}{a}\sinh(ax)$	
I9	$\frac{1}{x^2+a^2}$	$\frac{1}{a}\tan^{-1}\left(\frac{x}{a}\right)$	
I10	$\frac{1}{x^2-a^2}$	$\frac{1}{2a}\ln\left(\frac{x-a}{x+a}\right)$,	$\lvert x\rvert > a > 0$
I11	$\frac{1}{\sqrt{x^2-a^2}}$	$\cosh^{-1}\left(\frac{x}{a}\right)$,	$x > a > 0$
I12	$\frac{1}{\sqrt{a^2-x^2}}$	$\sin^{-1}\left(\frac{x}{a}\right)$,	$\lvert x\rvert < a$

Table 5.1. Table of standard integrals (a is a non-zero constant)

For more complicated integrands, such as products of functions, table 5.1 cannot be used immediately and we use a variety of techniques to transform the integrand a standard form. Subsequent sections describe these techniques.

As this chapter is a review of integration techniques we will merely illustrate some of the results in table 5.1 using the following examples. With practice it becomes unnecessary to write down the intermediate steps.

Problem 5.1

Carry out the following integrations using table 5.1 and the basic properties of integrals. Give your answer to (f) accurate to 4 decimal places.

(a) $\int x^4 - 2x^2 + 7\,dx$ (d) $\int \frac{1}{\sqrt{36-z^2}}\,dz$

(b) $\int \frac{1}{2}e^{3x} + e^{-x}\,dx$ (e) $\int_0^2 \frac{1}{t^2+4}\,dt$

(c) $\int \sin(4x) + 6\cos(3x)\,dx$ (f) $\int_4^6 \frac{1}{\sqrt{x^2-4}}\,dx$

Solution 5.1

(a) We integrate each term separately using I1 in table 5.1. Note that $\int k\,dx = kx$ corresponds to $n = 0$ in I1.

$$\begin{aligned}\int x^4 - 2x^2 + 7\,dx &= \int x^4\,dx - 2\int x^2\,dx + \int 7\,dx \\ &= \frac{1}{5}x^5 - \frac{2}{3}x^3 + 7x + c\end{aligned}$$

(b) Integration of exponentials is given by I3.

$$\int \frac{1}{2}e^{3x} + e^{-x}\,dx = \frac{1}{2}\int e^{3x}\,dx + \int e^{-x}\,dx = \frac{1}{6}e^{3x} - e^{-x} + c$$

(c) Trigonometric functions are given by I4, I5.

$$\begin{aligned}\int \sin(4x) + 6\cos(3x)\,dx &= \int \sin(4x)\,dx + \int 6\cos(3x)\,dx \\ &= -\frac{1}{4}\cos(4x) + 2\sin(3x) + c\end{aligned}$$

(d) There are a number of similar integrands but in this case it is I12.

$$\int \frac{1}{\sqrt{36-z^2}}\,dz = \int \frac{1}{\sqrt{6^2-z^2}}\,dz = \sin^{-1}\left(\frac{z}{6}\right) + c \qquad |z| < 6\,.$$

The condition $|z| < 6$ is necessary because the sine function can only return values between $+1$ and -1.
(e) In this example the integral is of the form I9 but limits are specified so we do not include the constant of integration.

$$\begin{aligned}\int_0^2 \frac{1}{t^2+4}\,dt &= \int_0^2 \frac{1}{t^2+2^2}\,dt = \left[\frac{1}{2}\tan^{-1}\left(\frac{t}{2}\right)\right]_0^2 \\ &= \frac{1}{2}\tan^{-1}(1) - \frac{1}{2}\tan^{-1}(0) = \frac{1}{2}\times\frac{\pi}{4} - 0 = \frac{\pi}{8}\end{aligned}$$

By convention we have taken the smallest positive values of the inverse tangent but of course we could have taken values of the form $\frac{\pi}{4} \pm n\pi$ and $\pm n\pi$ corresponding to the upper and lower limits, where n is an integer.
(f) This is of the type I11 and both limits satisfy the condition $|x| > 2$ which is necessary because $|\cosh u| \geq 1$ for all values of u.

$$\begin{aligned} \int_4^6 \frac{1}{\sqrt{x^2-4}}\, dx &= \int_4^6 \frac{1}{\sqrt{x^2-2^2}}\, dx = \left[\cosh^{-1}\left(\frac{x}{2}\right)\right]_4^6 \\ &= \cosh^{-1}(3) - \cosh^{-1}(2) = 0.4458 \end{aligned}$$

EXERCISES

Integrate the following.

5.1. $\int 2x - \sin(3x)\, dx$

5.2. $\int \frac{1}{2x^2+2}\quad dx$

5.3. $\int \frac{4}{u^5}\quad du$

5.4. $\int_0^1 6e^{3t}\, dt$

5.5. Find, to 3 decimal place accuracy, the value of x for which

$$\int_0^x 2e^{2t}\, dt = 1$$

5.6. Find b if

$$\int_0^1 \sqrt{\frac{3}{x}}\, dx = \int_0^b \sqrt{x}\, dx$$

5.3 Integration by substitution

This method is analogous to the chain rule in differentiation. To differentiate complicated functions it was advantageous to make a substitution in order to split the differentiation into simpler steps. For example, to differentiate

$$y = (3x^2 + 6)^4$$

we could make the substitution $u = 3x^2 + 6$ so that $y = u^4$. We would find $\frac{du}{dx}$ and $\frac{dy}{du}$ then substitute in the chain rule

$$\frac{dy}{dx} = \frac{dy}{du} \times \frac{du}{dx} \quad .$$

Substitution can also be a useful technique in integration. The aim is to make a change of variable so that the integral is transformed into one of the standard types of table 5.1. Ideally we would like to have the integral in the form

$$F(u) = \int f(u)\, du$$

where $f(u)$, the derivative of $F(u)$, is a standard function in table 5.1.

If u is a function of x, say $u = g(x)$, then

$$\frac{du}{dx} = g'(x),$$

and the chain rule for differentiation gives

$$\begin{aligned} \frac{d}{dx}F(u) &= \frac{d}{du}F(u) \times \frac{du}{dx} = f(u) \times \frac{du}{dx} \\ &= f(g(x)) \times g'(x) \, . \end{aligned}$$

We are interested in the reverse process, where an integral can be written in the form $\int f(g(x)) \times g'(x)\, dx$ so that the substitution $u = g(x)$ gives $du = g'(x)dx$ and $f(g(x)) = f(u)$. Hence

$$\int f(g(x)) \times g'(x)\, dx = \int f(u)\, du = F(u) \, .$$

The substitution can be either algebraic or trigonometric.

Algebraic substitution

We will illustrate the technique by example.

Problem 5.2

Use a substitution to find the following indefinite integral

$$\int \frac{1}{4x-3}\, dx \, .$$

Solution 5.2

If we let $u = 4x - 3$ we have the standard integral I2 in table 5.1. However, the integration is now with respect to u and not x. We have to find the link between du and dx.

We *differentiate* u to get

$$\frac{du}{dx} = 4 \text{ or } \frac{1}{4}\, du = dx \, .$$

We now rewrite the original problem in terms of u

$$\int \frac{1}{4x-3}\, dx = \int \frac{1}{u}\frac{1}{4}\, du = \frac{1}{4}\int \frac{1}{u}\, du = \frac{1}{4}\ln|u| + c \, .$$

Reverting to x gives the result

$$\int \frac{1}{4x-3}\,dx = \frac{1}{4}\ln|4x-3| + c\,.$$

This example illustrates how replacing ax with $ax+b$ would not significantly alter the results in table 5.1. If we make the substitution $u = ax+b$, b has no influence in $du = a\,dx$ and does not reappear until we replace u at the final step.

Problem 5.3

Use an appropriate substitution to determine the following integral

$$\int 24x(3x^2+6)^3\,dx\,.$$

Solution 5.3

Practice and experience helps in making the correct choice of substitution but a simple choice is to look for a function within a function, usually an obvious bracketed term. In this example we have $(3x^2+6)$ raised to a power so we try the substitution $u = 3x^2+6$.

Let $u = 3x^2+6$; then

$$\begin{aligned}
\frac{du}{dx} &= 6x \text{ or } \frac{1}{6}du = x\,dx,\\
\int 24x(3x^2+6)^3\,dx &= \int 24(3x^2+6)^3\,x\,dx\\
&= \int 4u^3\,du = u^4 + c = (3x^2+6)^4 + c\,.
\end{aligned}$$

Problem 5.4

By making an appropriate substitution evaluate the following integral to 3 decimal place accuracy:

$$\int_1^2 2te^{t^2-1}\,dt\,.$$

Solution 5.4

We *could* find the indefinite integral by means of a substitution and then impose the limits of integration. The obvious substitution is $u = t^2-1$ and hence $du = 2t\,dt$:

$$\int 2te^{t^2-1}\,dt = \int e^u\,du = e^u + c = e^{t^2-1} + c\,.$$

We now evaluate the definite integral

$$\int_1^2 2te^{t^2-1}\,dt = \left[e^{t^2-1}\right]_1^2 = e^3 - e^0 = 19.086\,.$$

In this example we used a substitution to find the indefinite integral and then reverted back to the original independent variable t before substituting the limits of integration.

An alternative is to change the limits of integration (when we make the substitution) from values of t to values of u and avoid the need to rewrite the integral as a function of t. The method is illustrated in the following problem.

Problem 5.5

Calculate the following integral accurate to 4 significant figures

$$\int_1^2 \frac{10t^2}{(t^3+1)^2}\,dt\,.$$

Solution 5.5

Let $u = t^3 + 1$; then $du = 3t^2\,dt$, and we now find the limits for u.
When $t = 1$, $u = 1^3 + 1 = 2$ and the lower limit is $u = 2$.
When $t = 2$, $u = 2^3 + 1 = 9$ and the upper limit is $u = 9$.
The integral becomes

$$\begin{aligned}\int_1^2 \frac{10t^2}{(t^3+1)^2}\,dt &= \frac{10}{3}\int_2^9 \frac{1}{u^2}\,du = \frac{10}{3}\int_2^9 u^{-2}\,du \\ &= \frac{10}{3}\left[-\frac{1}{u}\right]_2^9 = \frac{10}{3}\left(-\frac{1}{9}+\frac{1}{2}\right) = 1.296\,.\end{aligned}$$

Trigonometric substitutions

When differences or sums of squares appear in the integrand it is useful to make a trigonometric substitution and make use of trigonometric identities.

If $a^2 - x^2$ appears, use $x = a\sin\theta$ then $a^2(1-\sin^2\theta) = a^2\cos^2\theta\,.$
If $a^2 + x^2$ appears, use $x = a\tan\theta$ then $a^2(1+\tan^2\theta) = a^2\sec^2\theta\,.$
If $x^2 - a^2$ appears, use $x = a\sec\theta$ then $a^2(\sec^2\theta - 1) = a^2\tan^2\theta\,.$

Problem 5.6

Use the substitution $x = a\tan\theta$ to derive result I9 in table 5.1

$$\int \frac{1}{x^2+a^2}\,dx = \frac{1}{a}\tan^{-1}\left(\frac{x}{a}\right) + c\,.$$

Solution 5.6

The substitution leads to:

$$x = a\tan\theta, \qquad dx = a\sec^2\theta\,d\theta,$$

and the integral becomes

$$\begin{aligned}\int \frac{1}{x^2+a^2}\,dx &= \int \frac{1}{a^2(\tan^2\theta+1)}\,a\sec^2\theta\,d\theta \\ &= \int \frac{1}{(a\sec\theta)^2}\,a\sec^2\theta\,d\theta = \int \frac{1}{a}\,d\theta = \frac{\theta}{a} + c\,.\end{aligned}$$

But $\theta = \tan^{-1}\left(\frac{x}{a}\right)$, hence

$$\int \frac{1}{x^2+a^2}\,dx = \frac{1}{a}\tan^{-1}\left(\frac{x}{a}\right) + c\,.$$

EXERCISES

Use appropriate substitutions to determine the following integrals. Give your answers to 3 decimal place accuracy where appropriate.

5.7. $\int \frac{x}{x^2-1}\,dx$

5.8. $\int x(3x^2-1)^{\frac{1}{2}}\,dx$

5.9. $\int \frac{\sin x}{\cos^2 x}\,dx$

5.10. $\int_4^8 \frac{x}{\sqrt{x^2-15}}\,dx$

5.11. $\int_0^{\pi/2} \cos^2 x \sin x\,dx$

5.12. $\int_0^1 \frac{e^x}{(e^x+1)^2}\,dx$

5.13. $\int_0^{\pi/2} \frac{\cos x}{4+\sin^2 x}\,dx$

Use trigonometric substitutions to evaluate the following integrals.

5.14. $\int_0^1 \frac{1}{(1+x^2)^{3/2}}\,dx$

5.15. $\int_0^1 x^2\sqrt{1-x^2}\,dx$

5.4 Integration using partial fractions

5.4.1 Introduction

Prior to discussing the actual integration procedure it is necessary to consider briefly the manipulation of algebraic fractions.

In basic arithmetic we are often encouraged to combine fractions to obtain a neater result, for example,

$$\begin{aligned} \frac{1}{3}+\frac{1}{4} &= \frac{1\times 4}{3\times 4}+\frac{1\times 3}{4\times 3}=\frac{7}{12}, \\ \frac{3}{5}+\frac{2}{7} &= \frac{3\times 7}{5\times 7}+\frac{2\times 5}{7\times 5}=\frac{31}{35} . \end{aligned}$$

In each case we expressed the two fractions over a *common denominator* – the product of the individual denominators being convenient.

The same process can be applied to sums of algebraic fractions.

$$\begin{aligned} \frac{1}{x+1}+\frac{1}{x-2} &= \frac{x-2}{(x+1)(x-2)}+\frac{x+1}{(x+1)(x-2)}=\frac{2x-1}{(x+1)(x-2)} \quad \text{(A)} \\ \frac{1}{x-5}-\frac{2}{x+1} &= \frac{x+1}{(x-5)(x+1)}-\frac{2(x-5)}{(x-5)(x+1)}=\frac{11-x}{(x-5)(x+1)} \quad \text{(B)} \end{aligned}$$

Forming partial fractions is the reverse process, where a fraction whose denominator is the product of two or more factors is split into several simpler fractions.

At first this idea appears contrary to normal practice where we try to minimise the number of terms. The advantage becomes clear when integrating such algebraic functions. The terms on the far left of examples (A) and (B) integrate easily to $\ln|x+1|$, $\ln|x-2|$ and so on, but the combined terms on the right of examples (A) and (B) are not standard integrals.

Before describing the technique of partial fractions we first of all need to distinguish between proper and improper algebraic fractions (fractions whose denominator and numerator are polynomials).

Proper algebraic fractions

In arithmetic, for positive integers p and q, $\frac{p}{q}$ is a *proper* fraction if and only if $p < q$, e.g. $\frac{10}{13}$. If $p > q$ then $\frac{p}{q}$ is an *improper* fraction, e.g. $\frac{13}{10}$. In a similar way $\frac{p(x)}{q(x)}$ is a proper algebraic fraction, where $p(x)$ and $q(x)$ are polynomials, if and only if the *degree* of $p(x)$ is *strictly* less than the *degree* of $q(x)$. Otherwise we

have an improper algebraic fraction.
Examples of proper algebraic fractions:

$$\frac{3x+2}{4x^2-5x+2}, \quad \frac{R}{Rx+T}, \quad \frac{3x+2}{(x-4)(x+9)}, \quad \frac{2x^2-5}{(4x-1)(3x+2)(x-1)}.$$

Examples of improper algebraic fractions:

$$\frac{3x+2}{4x-5}, \quad \frac{2x^2+3x-5}{7x+11}, \quad \frac{3x^4}{x^2+2}, \quad \frac{7x^3}{(x-1)(x+2)(x-4)}.$$

An improper algebraic fraction can be rewritten as a leading polynomial plus a proper algebraic fraction. If the degree of $P(x)$ is greater than or equal to the degree of $q(x)$ in the expression $\frac{P(x)}{q(x)}$ then we express it in the form

$$\frac{P(x)}{q(x)} = \text{polynomial} + \frac{p(x)}{q(x)}$$

where $p(x)$ is at least one degree *less* than $q(x)$, i.e. $\frac{p(x)}{q(x)}$ is a proper algebraic fraction. The degree of the leading polynomial is the degree of $P(x)$ minus the degree of $q(x)$.

A procedure for converting an improper algebraic fraction into a proper one is illustrated in the next two examples. It should be noted that an alternative method would be to use algebraic long division.

Problem 5.7

Convert the following improper algebraic fraction into a form involving a proper algebraic fraction

$$\frac{4x-2}{2x+3}.$$

Solution 5.7

As $4x-2$ and $2x+3$ are of the same degree the leading polynomial is a constant (degree 0) and we have

$$\frac{4x-2}{2x+3} = A + \frac{B}{2x+3}$$

where A and B are constants.
Writing both terms on the right-hand side over the denominator $(2x+3)$

$$\frac{4x-2}{2x+3} = \frac{A(2x+3)+B}{2x+3}.$$

The denominators are the same on both sides, therefore we must also have equality of numerators

$$\begin{aligned} 4x - 2 &= A(2x+3) + B \\ \text{or} \qquad 4x - 2 &= 2Ax + 3A + B \ . \end{aligned}$$

This equation must hold for *all* values of x and matching the coefficients on the x terms and the constant terms gives

$$4 = 2A, \qquad\qquad -2 = 3A + B,$$

and so $A = 2$ and $B = -8$. Hence

$$\frac{4x-2}{2x+3} = 2 - \frac{8}{2x+3} \ .$$

Although not conclusive, a simple check is to substitute a particular value of x (the easiest being $x = 0$) and to compare both sides. In this example

$$\begin{aligned} \text{if } x = 0, \quad \frac{4x-2}{2x+3} &= -\frac{2}{3} \\ 2 - \frac{8}{2x+3} &= 2 - \frac{8}{3} = -\frac{2}{3} \end{aligned}$$

If the results match the *likelihood* is that the expressions are correct but if they differ it is *certain that* a mistake has been made.

Problem 5.8

Convert the following improper algebraic fraction into a form involving a proper algebraic fraction

$$\frac{x^4 - x + 7}{x^2 + 7} \ .$$

Solution 5.8

Since $x^4 - x + 7$ is two degrees higher than $x^2 + 7$ the leading polynomial is a quadratic and we can write

$$\frac{x^4 - x + 7}{x^2 + 7} = Ax^2 + Bx + C + \frac{Dx + E}{x^2 + 7} \ .$$

There are 5 constants to be determined: A, B and C appear as coefficients in the most general quadratic and then $Dx + E$ is the most general polynomial one degree less than the denominator $x^2 + 7$.

As before we write the terms on the right hand side over the same denominator and then equate the numerators:

$$\begin{aligned} x^4 - x + 7 &= (Ax^2 + Bx + C)(x^2 + 7) + Dx + E \,, \\ x^4 - x + 7 &= Ax^4 + Bx^3 + (C + 7A)x^2 + (7B + D)x + 7C + E \,. \end{aligned}$$

Matching the coefficients of powers of x:

$$1 = A,\ 0 = B,\ 0 = C + 7A,\ -1 = 7B + D,\ 7 = 7C + E,$$

which gives $A = 1$, $B = 0$, $C = -7$, $D = -1$ and $E = 56$. Hence

$$\frac{x^4 - x + 7}{x^2 + 7} = x^2 - 7 + \frac{56 - x}{x^2 + 7} \,.$$

EXERCISES

Express the following improper algebraic fractions as a polynomial plus a proper algebraic fraction.

5.16. $\frac{3x+2}{4x-5}$

5.17. $\frac{2x^2+3x-5}{7x+11}$

5.18. $\frac{3x^3-8x^2+5x-1}{3x+4}$

5.19. $\frac{3x^4}{x^2+2}$

5.20. $\frac{7x^3}{(x-1)(x+2)(x-4)}$

5.21. $\frac{4x^3+11x^2-9x-17}{(x-1)(x+2)}$

5.4.2 Resolving into partial fractions

We now describe the process of resolving a proper algebraic fraction into its partial fractions.

In general, the denominators of the partial fractions are the factors in the original fraction and the numerator of each partial fraction is one degree less than its denominator. Resolving an algebraic fraction whose denominator is the product of several factors will produce one partial fraction for each factor.

Problems 5.9 and 5.10 illustrate how to find the partial fractions of the following proper algebraic fractions.

$$\frac{13}{(x-2)(x+1)(2x-1)} \qquad \frac{x^2 + 4x - 16}{(x^2+1)(5x-6)}$$

The number of factors, and hence the number of partial fractions, in the denominators are 3 and 2 respectively. All the factors are linear in x except the

(x^2+1) factor in the second example. The partial fractions corresponding to the linear factors will be of the form $\frac{A}{x-2}, \frac{B}{x+1}$, etc. The partial fraction corresponding to (x^2+1) is of the form $\frac{Ax+B}{x^2+1}$, the numerator being the most general polynomial one degree less than the denominator.

Problem 5.9

Determine the following indefinite integral by first resolving the integrand into its partial fractions

$$\int \frac{13}{(x-2)(x+1)(2x-1)}\, dx\ .$$

Solution 5.9

The denominator has three linear factors so we are seeking three constants A, B and C such that

$$\frac{13}{(x-2)(x+1)(2x-1)} = \frac{A}{x-2} + \frac{B}{x+1} + \frac{C}{2x-1}\ .$$

Combining the fractions on the right-hand side over the common denominator $(x-2)(x+1)(2x-1)$ and equating the numerators gives

$$13 = A(x+1)(2x-1) + B(x-2)(2x-1) + C(x-2)(x+1)\ . \qquad (5.1)$$

There are two ways of finding the constants A, B and C. One method is to compare the coefficients of x and the constant term on each side of the equation,

$$13 = (2A+2B+C)x^2 + (A-5B-C)x + (2B-A-2C)\ .$$

Hence $0 = 2A+2B+C$, $0 = A-5B-C$ and $13 = 2B-A-2C$. Solving these linear equations gives $A = \frac{13}{9}$, $B = \frac{13}{9}$ and $C = -\frac{52}{9}$. An alternative method is to substitute particular values of x into equation (5.1) the important point to remember is that this equation must be true for *all* values of x. We can therefore choose to substitute values of x which make the bracketed terms on the right hand side zero and thus determine A, B and C directly.

$$\begin{aligned} \text{substitute } x = 2 \qquad 13 &= 9A + 0 + 0, \\ \text{substitute } x = -1 \qquad 13 &= 0 + 9B + 0, \\ \text{substitute } x = \frac{1}{2} \qquad 13 &= 0 + 0 - \frac{9}{4}C, \end{aligned}$$

and so $A = \frac{13}{9}$, $B = \frac{13}{9}$ and $C = -\frac{52}{9}$.

$$\frac{13}{(x-2)(x+1)(2x-1)} = \frac{13}{9(x-2)} + \frac{13}{9(x+1)} - \frac{52}{9(2x-1)}$$

as before. The integration can now be performed:

$$\begin{aligned}\int \frac{13}{(x-2)(x+1)(2x-1)}\,dx &= \frac{13}{9}\int \frac{1}{x-2} + \frac{1}{x+1} - \frac{4}{2x-1}\,dx \\ &= \frac{13}{9}\left(\ln|x-2| + \ln|x+1| - 2\ln|2x-1|\right) + c \\ &= \frac{13}{9}\ln\frac{|(x-2)(x+1)|}{(2x-1)^2} + c\,.\end{aligned}$$

Problem 5.10

Determine the following indefinite integral by first resolving the integrand into its partial fractions

$$\int \frac{x^2+4x-16}{(x^2+1)(5x-6)}\,dx\,.$$

Solution 5.10

There are two factors in the denominator so we will have two partial fractions, one of which will have a linear term in x as its numerator:

$$\frac{x^2+4x-16}{(x^2+1)(5x-6)} = \frac{Ax+B}{x^2+1} + \frac{C}{5x-6}\,.$$

As in the previous problem we write the fractions on the right hand side over the common denominator $(x^2+1)(5x-6)$ and then equate the numerators.

$$\begin{aligned} x^2+4x-16 &= (Ax+B)(5x-6) + C(x^2+1), \\ x^2+4x-16 &= (5A+C)x^2 + (5B-6A)x - 6B + C, \\ \text{giving } 1 = 5A+C, \quad 4 &= 5B-6A, \quad -16 = -6B+C\,. \end{aligned}$$

These three linear equations have the solution

$$A = 1, \quad B = 2, \quad C = -4\,.$$

and so the integral is

$$\begin{aligned}\int \frac{x^2+4x-16}{(x^2+1)(5x-6)}\,dx &= \int \frac{x+2}{x^2+1} - \frac{4}{5x-6}\,dx \\ &= \int \frac{x}{x^2+1} + \frac{2}{x^2+1} - \frac{4}{5x-6}\,dx \\ &= \frac{1}{2}\ln(x^2+1) + 2\tan^{-1}x - \frac{4}{5}\ln|5x-6| + c\,.\end{aligned}$$

EXERCISES

Find the following indefinite integrals by first resolving the integrand into its partial fractions.

5.22. $\int \frac{2x+3}{(x-1)(x-2)}\,dx$

5.23. $\int \frac{7}{(x-1)(x+2)}\,dx$

5.24. $\int \frac{5x+4}{x(x+5)}\,dx$

5.25. $\int \frac{-3}{x^2+2x-8}\,dx$

5.26. $\int \frac{-13x-19}{(x+1)(x-2)(x+3)}\,dx$

5.27. $\int \frac{8-8x-x^2}{x(x-1)(x-2)}\,dx$

5.28. $\int \frac{x^3+x^2+x+2}{(x^2+1)(x^2+2)}\,dx$

5.4.3 Repeated factors

So far, all the examples have been of the form

$$\frac{p(x)}{(\ldots)(\ldots)(\ldots)}$$

where the denominator comprised distinct linear and/or quadratic factors. We now consider repeated factors in the denominator. For example, x is a twice repeated factor in the expression.

$$\frac{1}{x^2(x+1)}\,.$$

The x^2 term is effectively a quadratic term, so if we are to remain consistent with our earlier examples we are seeking constants A, B and C such that

$$\frac{1}{x^2(x+1)} = \frac{Ax+B}{x^2} + \frac{C}{x+1}\,.$$

The first of the partial fractions can obviously be split to give

$$\frac{1}{x^2(x+1)} = \frac{A}{x} + \frac{B}{x^2} + \frac{C}{x+1}\,.$$

A similar process can be carried out for any repeated linear factors, for example

$$\begin{aligned}\frac{1}{(x+1)(x-1)^2} &= \frac{A}{x+1}+\frac{Bx+C}{(x-1)^2}\\ &= \frac{A}{x+1}+\frac{B(x-1)+(B+C)}{(x-1)^2}\\ &= \frac{A}{x+1}+\frac{B}{x-1}+\frac{C^*}{(x-1)^2}\end{aligned}$$

where $C^* = B + C$ is still an arbitrary constant to be determined.

Problem 5.11

Resolve the following algebraic fraction into its partial fractions

$$\frac{1}{(x+1)(x-1)^2}.$$

Solution 5.11

We could treat the repeated factor as a quadratic with a linear numerator, $Ax + B$, but it will reduce to the format already illustrated. We are trying to find three constants A, B and C such that

$$\frac{1}{(x+1)(x-1)^2} = \frac{A}{x+1}+\frac{B}{x-1}+\frac{C}{(x-1)^2}.$$

Writing the fractions on the right-hand side over the common denominator $(x+1)(x-1)^2$ and equating the numerators gives

$$1 = A(x-1)^2 + B(x+1)(x-1) + C(x+1)\,. \tag{5.2}$$

If we continue with our policy of substituting values of x which make the factors zero we will determine A and C but we will have to resort to comparing coefficients to determine B.

$$\begin{aligned}\text{Substitute } x=1: &\quad 1 = 2C,\\ \text{substitute } x=-1: &\quad 1 = 4A,\end{aligned}$$

and so $A = \frac{1}{4}$ and $C = \frac{1}{2}$. To find B we look at the coefficients of x^2 in equation (5.2). There is no x^2 term on the left-hand side, therefore $0 = A + B$. We already have $A = \frac{1}{4}$ and so $B = -\frac{1}{4}$.

$$\frac{1}{(x+1)(x-1)^2} = \frac{1}{4(x+1)} - \frac{1}{4(x-1)} + \frac{1}{2(x-1)^2}.$$

Problem 5.12

Carry out the following integration by first resolving the integrand into its partial fractions

$$\int \frac{x+1}{(2x-1)^2(x+3)}\,dx\,.$$

Solution 5.12

Again, we are trying to find three constants A, B and C such that

$$\frac{x+1}{(2x-1)^2(x+3)} = \frac{A}{2x-1} + \frac{B}{(2x-1)^2} + \frac{C}{x+3}\,.$$

Writing the terms on the right-hand side over the common denominator $(2x-1)^2(x+3)$ and equating the numerators gives

$$x+1 = A(2x-1)(x+3) + B(x+3) + C(2x-1)^2\,. \tag{5.3}$$

Substituting values of x which make the bracketed factors zero will determine B and C and by comparing coefficients we can determine A.

$$\begin{aligned} \text{Substitute } x = \frac{1}{2}: \qquad \frac{3}{2} &= \frac{7}{2}B, \\ \text{substitute } x = -3: \qquad -2 &= 49C, \end{aligned}$$

and so $B = \frac{3}{7}$ and $C = -\frac{2}{49}$. To find A compare the coefficients of x^2 in equation (5.3). There is no x^2 term on the left-hand side, therefore $0 = 2A + 4C$. We have already found that $C = -\frac{2}{49}$ and so $A = \frac{4}{49}$.

$$\begin{aligned} \int \frac{x+1}{(2x-1)^2(x+3)} &= \int \frac{4}{49(2x-1)} + \frac{3}{7(2x-1)^2} - \frac{2}{49(x+3)}\,dx \\ &= \frac{2}{49}\ln|2x-1| - \frac{3}{14(2x-1)} - \frac{2}{49}\ln|x+3| + c \\ &= \frac{2}{49}\ln\left|\frac{2x-1}{x+3}\right| - \frac{3}{14(2x-1)} + c\,. \end{aligned}$$

Problems 5.12 and 5.13 involved a twice repeated factor, the idea is extended for a factor repeated a greater number of times. For example, in the expression

$$\frac{11x-3}{(x+1)^3(x-6)^2}$$

the term $(x+1)$ is thrice repeated and $(x-6)$ is repeated twice. We would therefore seek the partial fractions

$$\frac{11x-3}{(x+1)^3(x-6)^2} = \frac{A}{x+1} + \frac{B}{(x+1)^2} + \frac{C}{(x+1)^3} + \frac{D}{x-6} + \frac{E}{(x-6)^2} \ .$$

When repeated factors occur, substituting specific values of x to make the factors zero will only give us the constant in the numerator above the ***highest power*** for each factor. To get the other constants we must compare coefficients.

EXERCISES

Find the following indefinite integrals by first expressing the integrand in terms of its partial fractions.

5.29. $\int \frac{3x+5}{(x+1)(x-1)^2} \ dx$

5.30. $\int \frac{4}{(x+1)^2(2x-1)} \ dx$

5.31. $\int \frac{x+1}{x^2(x-1)} \ dx$

5.32. $\int \frac{2x^3+5x^2-x+3}{(x-1)^2(x+2)^2} \ dx$

5.5 Integration by parts

The method of integration by parts is analogous to the product rule for differentiation:

$$\frac{d}{dx}(uv) = u\frac{dv}{dx} + v\frac{du}{dx} \ .$$

If we integrate this expression with respect to x we get

$$\begin{aligned} uv &= \int u\frac{dv}{dx}\,dx + \int v\frac{du}{dx}\,dx \\ \text{or} \quad \int u\frac{dv}{dx}\,dx &= uv - \int v\frac{du}{dx}\,dx \ , \\ \int uv'\,dx &= uv - \int vu'\,dx \ , \end{aligned}$$

where $u' = \frac{du}{dx}$ and $v' = \frac{dv}{dx}$. This provides a means of integrating simple products of functions, such as $x \ln x$ and $e^x \sin x$. The important step is making the right choice for u and v', noting that we differentiate u and integrate v'.

Sometimes there is no choice, as shown in the next example.

Problem 5.13

Use integration by parts to evaluate $\int \ln x \ dx$.

Solution 5.13

We are comparing this integral with

$$\int uv'dx = uv - \int vu'dx$$

and so we have to choose u and v' but it appears that there is only one term – we do not have a product. But, multiplying by unity, $\ln x = 1 \times \ln x$ and since we cannot integrate $\ln x$ directly we choose

$$u = \ln x \Rightarrow u' = \frac{1}{x}, \quad v = x \Rightarrow v' = 1.$$

Thus $$\int \ln x \, dx = x \ln x - \int 1 dx = x \ln x - x + c.$$

Problem 5.14

Use integration by parts to find the integral $\int x \cos x \, dx$.

Solution 5.14

We choose $u = x$ because it differentiates to give a constant and whether we differentiate or integrate $\cos x$ we will still end up with $\sin x$.

$$\text{So} \quad u = x, \qquad v' = \cos x,$$
$$u' = 1, \qquad v = \sin x.$$

Substituting these expressions into the integration by parts formula

$$\int x \cos x \, dx = x \sin x - \int \sin x dx = x \sin x + \cos x + c.$$

Note that we end up with a much easier integration on the right-hand side.

EXERCISES

Use the method of integration by parts to find the following integrals.

5.33. $\int \tan^{-1} x \, dx$

5.34. $\int \ln(x^2 + 4) \, dx$

5.35. $\int 3t^2 e^t \, dt$

5.36. $\int 10e^{-6t} \sin(2t) \, dt$

5.37. $\int_0^1 xe^x \, dx$

5.38. Calculate to three decimal places.

$$\int_1^4 \ln x \, dx$$

EXERCISES

5.39. Evaluate

$$\int_3^\infty \frac{1}{x^2}\,dx\,.$$

5.40. Use integration to calculate the area bounded by the curve $y = -x^2$ and the x–axis between $x = 3$ and $x = 6$. Comment on the sign of the integral.

5.41. Find the magnitude of the finite area enclosed between the curves

$$y = 2 - x^2 \quad \text{and} \quad y - x = -4\,.$$

5.42. If the *area* under the curve $y = f(x)$ between $x = a$ and $x = b$ is rotated about the x–axis then a volume is produced. This is called a *solid of revolution* and the volume can be found by integration. When a narrow strip of width Δx is rotated about the x–axis a disc is generated of radius y and thickness Δx and its volume is $\pi y^2 \Delta x$. If we sum all such discs between $x = a$ and $x = b$ we get

$$V = \lim_{\Delta x \to 0} \sum \pi y^2 \Delta x = \int_a^b \pi y^2\,dx\,.$$

Calculate the volume of the solid of revolution obtained by rotating the curve $y = \mathrm{e}^{-x}$ through 360° about the x–axis between $x = 0$ and $x = 4$. Give your answer in cubic centimeters to 3 decimal places.

5.43. When finding a Fourier series to represent a function $f(x)$ over the range $-\pi < x < \pi$ it is necessary to evaluate the Fourier coefficients a_n and b_n (n a positive integer) which are given by

$$\begin{aligned} a_n &= \frac{1}{\pi}\int_{-\pi}^{\pi} f(x)\cos(nx)\,dx, \\ b_n &= \frac{1}{\pi}\int_{-\pi}^{\pi} f(x)\sin(nx)\,dx\,. \end{aligned}$$

Use integration by parts to find a_n and b_n when $f(x) = x$.

5.44. If r is a constant, integrate the following.

$$\int t r^t\,dt$$

Hint: $r^t = \mathrm{e}^{\ln r^t}$.

6
Applications of Differentiation

6.1 Introduction

It is easy to conveniently forget or gloss over the underlying conditions and theory for the existence of maxima and minima of functions. In terms of 'the ends justify the means' this is perhaps understandable, but it is important to have a good grasp of the background material; this is covered in section 6.2.

This is followed by sections on the Mean Value Theorem and the sketching of curves. A common application of differentiation is the optimisation of functions subject to constraints. This is covered in section 6.4.

The final section on Taylor series shows how repeated differentiation about a point can generate polynomials which approximate other functions, such as exponentials.

6.2 Functions

6.2.1 Increasing and decreasing functions

In chapter 4 we defined the rate of change of a function $f(x)$ to be the derivative $f'(x)$ which could also be interpreted as the gradient of the tangent to the curve. The general features are

1. if $f'(x_0) < 0$ (negative gradient) then f is *decreasing* at $x = x_0$
2. if $f'(x_0) > 0$ (positive gradient) then f is *increasing* at $x = x_0$
3. if $f'(x_0) = 0$ (zero gradient) then f is *stationary* at $x = x_0$

A simple function which exhibits all of these features is $f(x) = \sin x$ on the interval $[0, 2\pi]$, as illustrated in figure 6.1. The intervals over which $f(x)$ is

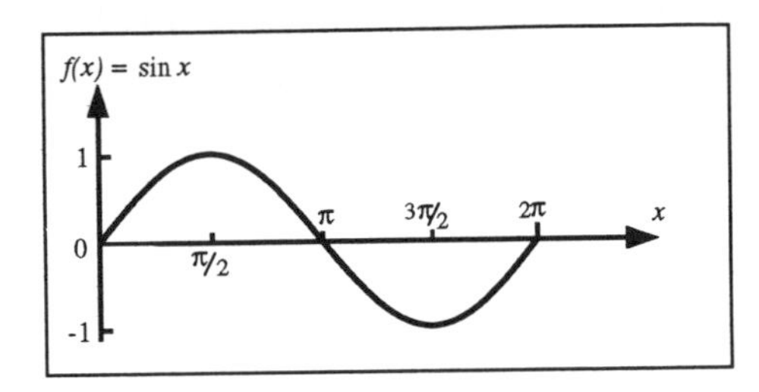

Fig. 6.1. Graph of $f(x) = \sin x$

increasing or decreasing can be deduced by looking at the graph of the function or by analysing its derivative $f'(x) = \cos x$.

By inspection, the function appears to be increasing on the intervals $[0, \frac{\pi}{2})$ and $(\frac{3\pi}{2}, 2\pi]$. The function is decreasing on the interval $(\frac{\pi}{2}, \frac{3\pi}{2})$ and has maximum and minimum values at the points $x = \frac{\pi}{2}$ and $x = \frac{3\pi}{2}$ respectively.

The same conclusions could be deduced by finding those values of x for which $f'(x)$ is positive (f increasing), negative (f decreasing) and zero (f stationary).

The idea of increasing and decreasing functions is intuitively obvious. The notion of a *stationary* point can be explained as follows: *if* a ball is placed precisely at one of these points then it will not move. If the ball is placed on a section of the curve which is increasing then the ball will roll back down the slope and if placed on a decreasing section of the curve it will roll forward.

The advantages of using the derivative to determine when a function is increasing or decreasing are twofold. The first is precision; we can only *estimate* coordinates visually. The second is time; it is usually quicker to differentiate a function and solve the resulting equation than to generate a neat sketch of a curve. However, curve sketching (section 6.3) is an important tool in its own right because it gives us an overall view of the behaviour of a function.

6.2.2 Relative maxima and minima

The graphs of most functions have 'hills' and 'valleys'; the graph of $\sin x$ in figure 6.1 had one of each but if we had plotted the curve for $0 \leq x \leq 6\pi$ there would have been three hills and three valleys.

The peaks of the hills and the bottoms of valleys are referred to as *relative* maxima and minima to indicate they are the highest or lowest points in the vicinity or neighbourhood.

Definition 6.1

A function f is said to have a *relative maximum* at x_0 if $f(x_0) \geq f(x)$ for all x in some open interval containing x_0.
It is important that the interval is *open* as it then includes points either side of x_0.
A function f is said to have a *relative minimum* at x_0 if $f(x_0) \leq f(x)$ for all x in some open interval containing x_0.
A function is said to have a *relative extremum* at x_0 if it has either a relative maximum or a relative minimum at x_0.

Definition 6.2

A *critical point* of a function is any value of x in the domain of f for which either $f'(x) = 0$ (stationary point) or where f is not differentiable.
In most cases the critical points will be those at which $f'(x) = 0$ but some functions are not differentiable at particular points. For example, at $x = 0$ $f(x) = \frac{1}{x}$ is not defined and therefore not differentiable.

First derivative test

We can determine the nature of critical points using the first derivative test which looks at the change in *sign* of the first derivative as x increases through x_0.

1. $f(x)$ has a relative maximum at x_0 if $f'(x)$ changes sign from positive to negative.
2. $f(x)$ has a relative minimum at x_0 if $f'(x)$ changes sign from negative to positive.
3. $f(x)$ has neither a relative maximum nor a relative minimum at x_0 if $f'(x)$ does not change sign. Such a point is called a point of inflection.

The above descriptions mirror the behaviour of the tangent to the graph of the function. Just before a maximum the tangent has a positive gradient which then decreases to zero and subsequently becomes negative after passing through the maximum.

Problem 6.1

Find and classify the critical points of

$$f(x) = 2x^3 - 3x^2 - 12x + 13 .$$

Solution 6.1

There are no points at which f is not differentiable. Therefore the only critical points are those where $f'(x) = 0$, i.e.

$$f'(x) = 6x^2 - 6x - 12 = 6(x-2)(x+1) = 0 \; .$$

Thus $x = 2$ and $x = -1$. The corresponding function values are -7 and 20 respectively and the critical points are $(-1, 20)$ and $(2, -7)$.
These critical points split the x-axis into three intervals: $-\infty < x < -1$, $-1 < x < 2$ and $2 < x < \infty$. We know the sign of $f'(x)$ only changes at a critical point, therefore we can use a single test value in each of the intervals.

Interval	$-\infty < x < -1$	$-1 < x < 2$	$2 < x < \infty$
Test value of x	-2	0	3
Sign of $f'(x)$	$+$	$-$	$+$

Hence $(-1, 20)$ is a relative maximum and $(2, -7)$ is a relative minimum.

6.2.3 Concavity

An arc of a curve is called *concave down* if it lies below its tangent line and *concave up* if it lies above its tangent line, see figure 6.2. Notice that the slope

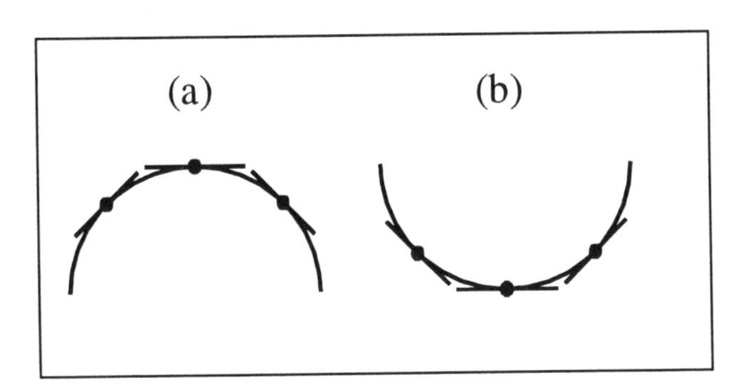

Fig. 6.2. Concave (a) down (b) up

of the tangent decreases (moving from left to right) along a curve which is concave down and the slope of the tangent increases along a curve which is concave up.

The slope of the tangent is the first derivative, $f'(x)$, and the *change* in slope of the tangent is the second derivative, $f''(x)$. The following definition states formally what we have observed.

Definition 6.3

If $f''(x) < 0$ on an open interval then $f(x)$ is *concave down* on the interval.
If $f''(x) > 0$ on an open interval then $f(x)$ is *concave up* on the interval.

Problem 6.2

Find open intervals for which $f(x) = 2x^3 - 3x^2 - 12x + 13$ is (a) concave up and (b) concave down.

Solution 6.2

We need the second derivative

$$\begin{aligned} f(x) &= 2x^3 - 3x^2 - 12x + 13, \\ f'(x) &= 6x^2 - 6x - 12, \\ f''(x) &= 12x - 6 . \end{aligned}$$

The second derivative is zero when $x = \frac{1}{2}$, it is positive when $x > \frac{1}{2}$ and negative when $x < \frac{1}{2}$. Therefore the function is (a) concave up on the interval $\frac{1}{2} < x < \infty$ and (b) concave down on the interval $-\infty < x < \frac{1}{2}$.

Second derivative test

From figure 6.2 we see that a relative maximum occurs when $f'(x) = 0$ *and* the function is concave down. Similarly, a relative minimum occurs when $f'(x) = 0$ *and* the function is concave up. These observations provide the basis of the second derivative test:

1. $f(x)$ has a relative maximum at x_0 if $f''(x_0) < 0$
2. $f(x)$ has a relative minimum at x_0 if $f''(x_0) > 0$
3. If $f''(x_0) = 0$ or does not exist then the second derivative test *fails* and the first derivative test should be used.

We can redo problem 6.1 using the second derivative test. The critical points were found to be $(-1, 20)$ and $(2, -7)$ and the second derivative is

$$f''(x) = 12x - 6 \quad .$$

Substituting the values of x we find

$$x = -1 : \quad f''(x) = -18 \ , \quad x = 2 : \quad f''(x) = 18 \ .$$

The numeric values are of secondary importance, we are interested in the *sign* of the second derivative. The second derivative test indicates that $(-1, 20)$ is a relative maximum while $(2, -7)$ is a relative minimum.

A slight shortcoming of this test is that we have to remember whether a positive second derivative means a relative minimum or a maximum, and similarly for a negative second derivative. It is useful to think of the functions $y = x^2$ and $y = -x^2$. These functions are easy to visualise as having a minimum and a maximum respectively and their second derivatives are $+2$ and -2.

6.2.4 Points of inflection

A point of inflection is a point where the concavity of a curve changes from up to down or vice versa.

Definition 6.4

A curve has a point of inflection at x_0 if $f''(x_0) = 0$ or $f''(x_0)$ is not defined *and* if $f''(x)$ changes sign as x increases through x_0. If also $f'(x_0) = 0$ then it is a stationary point of inflection.

The emphasis on the '*and*' is important because both conditions must be satisfied. For example, if $f(x) = x^3$ we find $f''(x) = 6x = 0$ at the point $x = 0$ and this turns out be a point of inflection because the curve *also* changes from concave down to concave up as x increases through 0.

However, if $f(x) = x^4$ then $f''(x) = 4x^3 = 0$ at $x = 0$ but this is not a point of inflection because the concavity does not change. In fact this point is a minimum.

6.2.5 Absolute maxima and minima

So far we have considered *relative* (local) maxima and minima; now we investigate *absolute* (global) maximum and minimum values. Consider figure 6.3 where the *continuous* function $f(x) = 4 - x^2$ has been drawn over three intervals.

In (a) the domain is the *closed* interval $[-2, 1]$ and the function has an *absolute minimum* value of zero at $x = -2$ and an *absolute maximum* value of 4 at $x = 0$.

In (b) the domain is now open at $x = -2$ and the function has the same *absolute maximum* value of 4 at $x = 0$. There is no *absolute minimum* value because x never quite attains the value -2. However close a value to -2 we take there will always be another value closer to -2 for which $f(x)$ is smaller.

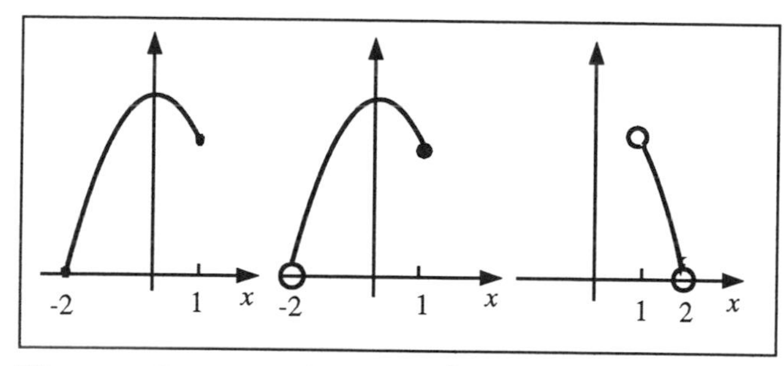

Fig. 6.3. Graphs of $f(x) = 4 - x^2$ on different intervals

In (c) the domain is the *open* interval $(1, 2)$ and the function has neither an *absolute minimum* nor an *absolute maximum* value on the interval, for the same reason as in (b).

Discontinuous functions require more care. Perhaps the most common discontinuous function is $f(x) = \tan x$. This is discontinuous whenever x is an odd multiple of $\frac{\pi}{2}$ and so there are no absolute maximum or minimum values when the domain of x includes an odd multiple of $\frac{\pi}{2}$.

The preceding paragraphs indicate that the existence of absolute maximum and minimum values depends on the type of interval and on the continuity or otherwise of the function. This is summarised in the following definition.

Definition 6.5

If a function is *continuous* on a closed interval $[a, b]$ then f has an absolute maximum and an absolute minimum value on $[a, b]$. These absolute values occur either at a critical point or at an endpoint of the interval.

Rolle's theorem

If $f(x)$ is continuous in the closed interval $[a, b]$ and differentiable in the open interval (a, b) and if $f(a) = f(b)$ then there exists at least one point x_0 in (a, b) such that $f'(x_0) = 0$.

Proof

If $f(x) = f(a)$ everywhere in (a, b) then $f'(x) = 0$ everywhere and the theorem is proved. If this is not the case then there is a point $x_0 \in (a, b)$ at which $f(x)$ takes its maximum value, i.e. $f(x) \leq f(x_0)$ for every point in the interval.

Therefore, for any small value h it must be true that $f(x_0) - f(x_0 + h) \geq 0$. If h is positive

$$\frac{f(x_0 + h) - f(x_0)}{h} \leq 0$$

and if we let $h \to 0$ we obtain $f'(x_0) \leq 0$.

Similarly, if h is negative then

$$\frac{f(x_0+h)-f(x_0)}{h} \geq 0$$

and if we let $h \to 0$ we obtain $f'(x_0) \geq 0$.
This combination of inequalities gives $f'(x_0) = 0$ and the theorem is proved.□

The conditions that the function is continuous and differentiable allow us to visualise the result using the tangent to the graph of the function. If $f'(x_0) = 0$ then the tangent to the curve is horizontal at the point x_0. Rolle's theorem states there is *at least* one point x_0; figure 6.4 shows two possible scenarios.

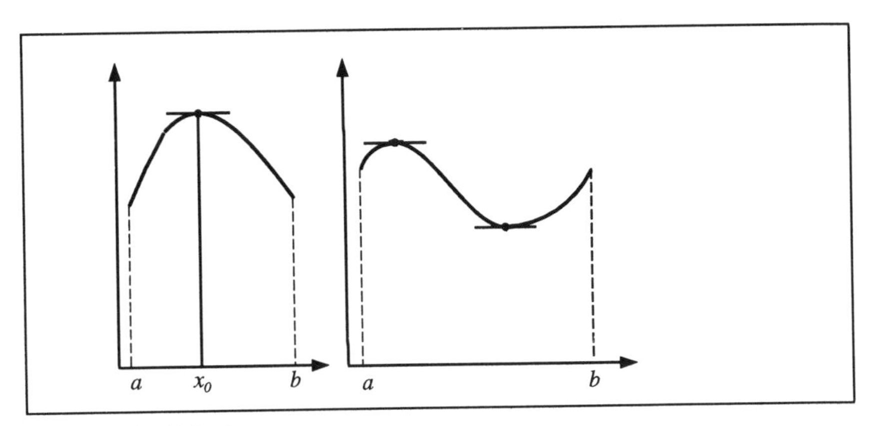

Fig. 6.4. Rolle's theorem

Problem 6.3

Find a value for x_0 as prescribed by Rolle's theorem, given

$$f(x) = \sin x \ , \qquad 0 \leq x \leq \pi.$$

Solution 6.3

The function is continuous and $f(0) = f(\pi) = 0$ so the conditions on Rolle's theorem are satisfied. Hence we can find the point x_0.

$$f(x) = \sin x \ \Rightarrow \ f'(x) = \cos x$$

$$\text{If} \quad f'(x_0) = \cos x_0 = 0 \qquad \text{then} \quad x_0 = \frac{\pi}{2}\ .$$

The Mean Value Theorem

If $f(x)$ is continuous in the closed interval $[a, b]$ and differentiable in the open interval (a, b) then there exists at least one point x_0 in (a, b) such that

$$f'(x_0) = \frac{f(b) - f(a)}{b - a} .$$

That is, there exists at least one point in the interval (a, b) where the gradient of the tangent is equal to the gradient of the straight line joining the points $(a, f(a))$ and $(b, f(b))$, as in figure 6.5.

The Mean Value Theorem (MVT) is an extension of Rolle's theorem. The straight line joining the points $(a, f(a))$ and $(b, f(b))$ is called the *secant line.*

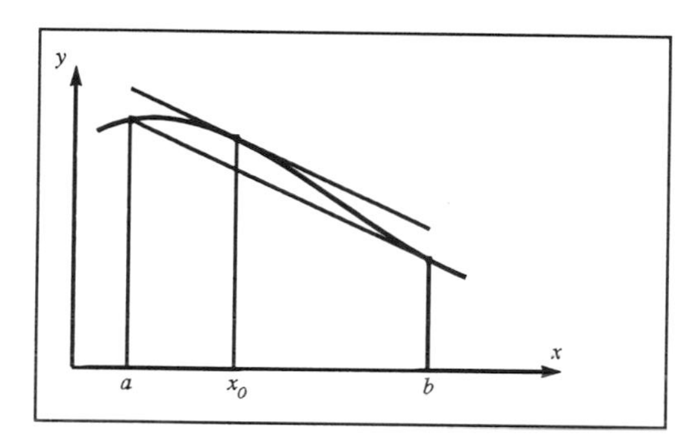

Fig. 6.5. Illustration of the MVT

Proof

We would like to use Rolle's theorem immediately but the function probably has different values at the endpoints of the interval, $x = a$ and $x = b$.

First we find the equation of the secant line. Knowing its gradient and that it passes through the point $(a, f(a))$ we can write

$$y = f(a) + \frac{f(b) - f(a)}{b - a}(x - a) .$$

We define $F(x)$ to be the *difference* of $f(x)$ and the secant line, y.

$$F(x) = f(x) - y = f(x) - f(a) - \frac{f(b) - f(a)}{b - a}(x - a) .$$

Notice now that $F(a) = F(b) = 0$. Also, $F(x)$ satisfies the conditions in Rolle's theorem whenever $f(x)$ does.

Hence there exists at least one point x_0 in (a, b) such that $F'(x_0) = 0$. Differentiating we have

$$F'(x) = f'(x) - \frac{f(b) - f(a)}{b - a},$$

and therefore at some point x_0

$$f'(x_0) - \frac{f(b) - f(a)}{b - a} = 0$$

which gives the result

$$f'(x_0) = \frac{f(b) - f(a)}{b - a}$$

and the theorem is proved.□

This result can be rearranged to give

$$f(b) = f(a) + (b - a)f'(x_0) .$$

Or, if we write $b = a + h$ and $c = x_0$ we have the form

$$f(a + h) = f(a) + hf'(c), \qquad c \in (a, a + h) .$$

The following problems show some applications of the MVT.

Problem 6.4

Find a value for x_0 as prescribed by the MVT, given that

$$f(x) = x^2 - 3x + 4 \ , \qquad 1 \leq x \leq 3.$$

Solution 6.4

All polynomials are continuous and differentiable, therefore the conditions for the MVT are satisfied. The values of f at the end points are

$$f(1) = 2 \quad \text{and} \quad f(3) = 4 .$$

Differentiating the function and applying the MVT gives

$$\begin{aligned} f'(x) &= 2x - 3, \\ f'(x_0) = 2x_0 - 3 &= \frac{f(3) - f(1)}{3 - 1}, \\ 2x_0 - 3 &= \frac{4 - 2}{3 - 1} = 1 \Rightarrow x_0 = 2 . \end{aligned}$$

Problem 6.5

Two stationary police patrol cars equipped with radar are 5 miles apart on a motorway. As a motorcycle passes the first patrol car its speed is registered at 55 mph. Four minutes later, when it passes the second patrol car, its speed registers 50 mph.

Prove that the motorcycle must have exceeded the speed limit of 70 mph at some time between the two patrol cars.

Solution 6.5

Let the independent variable be time t measured in hours and the function be the distance $s(t)$ travelled by the motorcycle measured in miles from the first police patrol car. Four minutes is $\frac{1}{15}$ of an hour and so

$$s(0) = 0 \qquad\qquad s(1/15) = 5\ .$$

Applying the MVT on the time interval $(0, \frac{1}{15})$ there exists a t_0

$$s'(t_0) = \frac{s(\frac{1}{15}) - s(0)}{\frac{1}{15} - 0} = \frac{5-0}{\frac{1}{15} - 0} = 75\ .$$

But $s'(t_0)$ is the speed of the motorcycle at time t_0 and so at some time t_0 the speed of the motorcycle is 75 mph, i.e. the speed limit is exceeded.

Problem 6.6

Use the MVT to prove $\tan x > x\ ,\quad 0 < x < \frac{\pi}{2}$.

Solution 6.6

Consider any interval $[0, b]$ where $0 < b < \frac{\pi}{2}$. On such an interval $f(x) = \tan x$ is continuous and differentiable so the MVT implies that there exists a value x_0 in the interval $(0, b)$ such that

$$\begin{aligned} \tan b &= \tan 0 + (b-0)f'(x_0) \\ \tan b &= b\sec^2 x_0 \end{aligned}$$

But $\quad \sec^2 x_0 = \dfrac{1}{\cos^2 x_0} > 1\ ,\qquad x_0 \in (0, b)\ \text{and}\ b < \dfrac{\pi}{2}$

Hence $\tan b > b$, for all $b \in (0, \frac{\pi}{2})$ and we have the required result.

If we consider the MVT in the form $f(a+h) = f(a) + hf'(c)$ we see that $f(a)$ can be regarded as an approximation to $f(a+h)$, the error term in this approximation being $hf'(c)$. The following problem shows how the MVT can be used to estimate the error involved when making such approximations.

Problem 6.7

Use the MVT to bound the error in approximating $(4.1)^{-1/2}$ by the value $4^{-1/2} = 0.5$. Compare this error bound with the actual error given that $(4.1)^{-1/2} = 0.493865$ to 6 decimal places.

Solution 6.7

If we define $f(x) = x^{-1/2}$, $a = 4$, $h = 0.1$ then f satisfies the conditions of the MVT in the interval $(4, 4.1)$.
Therefore there exists a value c in the interval $(4, 4.1)$ such that

$$\begin{aligned} (4.1)^{-1/2} &= 4^{-1/2} + 0.1 f'(c) \\ (4.1)^{-1/2} &= 4^{-1/2} + 0.1 \left(-\frac{1}{2} c^{-3/2} \right) \\ &= 0.5 - \frac{0.05}{c^{3/2}}, \qquad 4 < c < 4.1 \, . \end{aligned}$$

Thus $(4.1)^{-1/2} \approx 0.5$ with error $\frac{-0.05}{c^{3/2}}$.
Notice the error has maximum size (in absolute value) when $c = 4$ in the denominator.
Hence

$$|\text{error}| < \left| \frac{-0.05}{4^{3/2}} \right| = 0.00625 \, .$$

The modulus of the actual error is 0.006135 to 6 decimal places.

EXERCISES

6.1. Find a value for x_0 as prescribed by the MVT given that

$$y = \ln x, \qquad 1 \leq x \leq \mathrm{e} \, .$$

6.2. Use the MVT to show that $\sin x < x$ for $x > 0$.

6.3. If T is the period of oscillation of a simple pendulum of length l then

$$T = 2\pi \sqrt{\frac{l}{g}} = f(l)$$

where the constant g is the acceleration due to gravity. If, to keep the correct time, a pendulum should be exactly of length l but due to expansion or faulty measurement by its maker its actual length is $l + \epsilon$, where ϵ can be positive or negative, use the MVT to estimate the corresponding error in the period.

6.3 Curve sketching

A sketch means a neat, approximate graph that clearly indicates the general shape of the curve and highlights important points such as crossing of axes, relative maxima/minima, points of inflection, discontinuities, etc.

In this section we shall illustrate the procedure by sketching polynomials, but the principles are the same for more complex functions.

Consider the curve in figure 6.6. Imagine that you are trying to describe this curve verbally in order that someone who is unsighted can reproduce it.

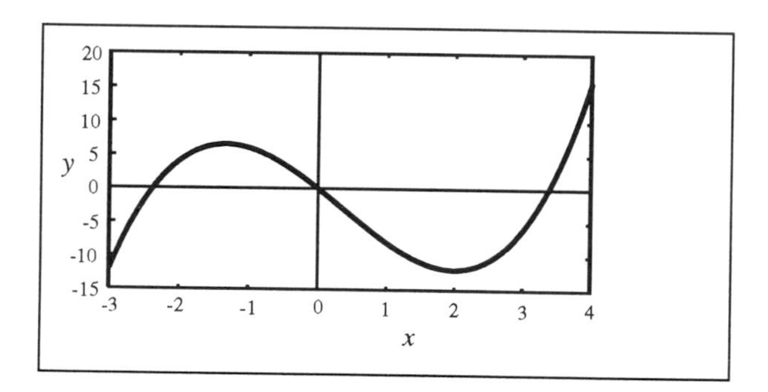

Fig. 6.6. Graph of an arbitrary function

Where do you start?

The first things to be drawn are the coordinate axes – so first define the intervals for x and y.

What next?

List all the important points: where it crosses the axes, where the turning points are and their nature, any discontinuities, etc.

Having followed these two steps and marked the important points it is simply a case of joining the points neatly to make a smooth curve.

How do we find the points of most interest?

The preceding sections of this chapter have shown how to find relative extrema and points of inflection. The points where the curve crosses the coordinate axes usually involve solving simple equations.

In the case of polynomials the sketching can be limited to low powers of x because this is sufficient to illustrate most types of behaviour.
A set of general rules is as follows.

1. Where are the axes crossed (if at all)?
   A polynomial of degree n can cross the y-axis once and the x-axis *at most* n times.
2. Find and classify the relative extrema.
3. Find any points of inflection.
4. If a *bounded* interval is specified for x find the endpoints of the curve.

Find all of these points *before* drawing the coordinate axes so that appropriate scaling can be chosen.

Problem 6.8

Locate any turning (stationary) points and their nature for the function

$$y = 2x^3 - 3x^2 - 12x + 13 = (x-1)(2x^2 - x - 13) \ .$$

By determining where the curve crosses the coordinate axes sketch the curve over the interval $-3 \le x \le 4$.

Solution 6.8

Most of the important points have already been found in the problems of this chapter but we shall use the suggested procedure.
Axis crossing:
The curve crosses the y-axis when $x = 0$, i.e. $y = 13$.
The curve crosses the x-axis when $y = 0$,

$$(x-1)(2x^2 - x - 13) = 0 \ .$$

This equation is satisfied by $x = 1$ and by the roots of the quadratic

$$x = \frac{1 \pm \sqrt{105}}{4} \quad \Rightarrow \quad x = 2.812 \ , \ x = -2.312 \ .$$

For a sketch the values $x = 2.8$ and $x = -2.3$ will be sufficiently accurate.
Relative extrema:
Stationary points are where $\frac{dy}{dx} = 0$:

$$\frac{dy}{dx} = 6x^2 - 6x - 12 = 6(x-2)(x+1) = 0,$$

giving the solutions $x = 2$ and $x = -1$. The corresponding y values are -7 and 20 respectively and so the stationary points are at $(-1, 20)$ and

$(2,-7)$. We determine their nature using the second derivative test. The second derivative is

$$\frac{d^2y}{dx^2} = 12x - 6$$

and substituting the values of x gives

$$x = -1: \quad \frac{d^2y}{dx^2} = -18\,, \quad x = 2: \quad \frac{d^2y}{dx^2} = 18\,.$$

The second derivative test indicates that $(-1,20)$ is a relative maximum while $(2,-7)$ is a relative minimum.

Points of inflection:

These occur when the second derivative is zero

$$\frac{d^2y}{dx^2} = 12x - 6 = 0 \Rightarrow x = \frac{1}{2},\ y = \frac{13}{2}\,.$$

End-points:

The end-points are

$$x = -3\,,\ y = -32 \qquad x = 4\,,\ y = 45\,.$$

Having found all the points of interest we can sketch the curve, noting that $-3 \le x \le 4$ and $-32 \le y \le 45$.

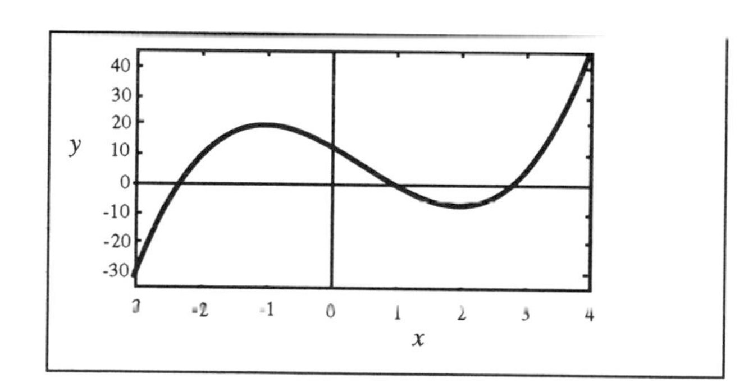

Fig. 6.7. Sketch of $y = 2x^3 - 3x^2 - 12x + 13$

EXERCISES

6.4. Find the turning points of $y = x^3 - 2x + 1$ and determine their nature.
Given that the curve crosses the x-axis when $x = 1$ find any other x-axis crossings and hence sketch the curve for $-2 \le x \le 3$.

6.5. Find the turning points of $y = x^4 - 8x^2 + 16 = (x^2 - 4)^2$ and determine their nature.
Sketch the curve for $-3 \le x \le 3$.

6.4 Optimisation

The most common application of differentiation is the optimisation of a function subject to constraints. We shall consider problems where a function of two variables is to be maximised or minimised subject to one constraint. In such problems the constraint is used to eliminate one of the variables, leaving a function of only *one* variable.

Problem 6.9

If a and b are two non-negative numbers prove that when $a+b=10$ the minimum value of a^2+b^2 is 50.

Solution 6.9

We are given the constraint $a+b=10$ and want to minimise the function $y=a^2+b^2$. We use the constraint to write b in terms of a and so reduce y to a function of only one variable.
We write $b=10-a$ and so

$$\begin{aligned} y &= a^2+(10-a)^2 = 2a^2-20a+100, \\ \frac{dy}{da} &= 4a-20\,, \qquad \frac{d^2y}{da^2}=4\,. \end{aligned}$$

For a maximum or a minimum $\frac{dy}{da}$ must be zero

$$0=4a-20 \Rightarrow a=5\,,\ b=10-a=5\,.$$

This is a minimum because the second derivative is positive. Hence the minimum value is $y=a^2+b^2=5^2+5^2=50$.

In some cases there may be more than one solution but we can usually reject some of these on practical grounds, as shown in the next problem.

Problem 6.10

A rectangular page is to contain 24 square inches of print. The margins at the top and bottom of the page are each 1.5 inches. The margins on each side are 1 inch. If y is the height of the printed area, show that A, the total area of the page, is given by

$$A=30+2y+\frac{72}{y}$$

Find the dimensions of the *page* which use the least amount of paper.

Solution 6.10

When a problem involves geometric constraints it is advisable to draw a sketch indicating the relevant features and dimensions. Let y be the

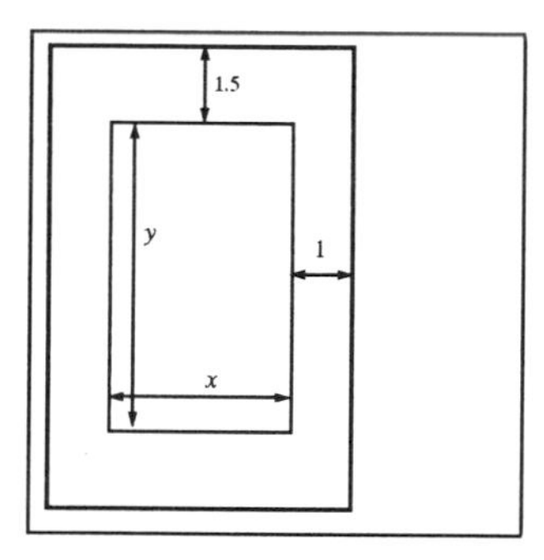

Fig. 6.8. Printed page

height of the printed area and x be the width of the printed area. The constraint is $xy = 24$. The height of the page is $y + 3$ i.e. y plus the top and bottom margins, and similarly the width of the page is $x + 2$, as indicated in figure 6.8. Hence the area of the page is $A = (y + 3)(x + 2)$ and we use the constraint to express x in terms of y, $x = \frac{24}{y}$. The area is

$$A = (y + 3)(x + 2) = (y + 3)\left(\frac{24}{y} + 2\right) = 30 + 2y + \frac{72}{y} .$$

The area is now a function of only one variable and we can find the critical points by differentiating and setting the derivative to zero.

$$\frac{dA}{dy} = 2 - \frac{72}{y^2} = 0 \Rightarrow y = \sqrt{36} = \pm 6 .$$

There are two solutions to the *mathematical* problem but we take the positive value as the solution to the *physical* problem because a negative length is meaningless in this context. The height and width of the page are $y + 3 = 9$ inches and $x + 2 = \frac{24}{y} + 2 = 6$ inches respectively.
We have found a critical point but do not yet know if it is a maximum or minimum. This is deduced from the sign of the second derivative:

$$\text{at } y = 6, \qquad \frac{d^2A}{dy^2} = \frac{144}{y^3} = \frac{144}{6^3}$$

which is positive, hence we have found a minimum.

EXERCISES

6.6. The sum of two non-negative numbers is 28. Find the two numbers if the cube of one number plus the square of the other is as small as possible.

6.7. The sum of the lengths of the two shortest sides of a right-angled triangle is 20 centimetres. What is the maximum possible area of the triangle?

6.8. A cylindrical can to hold 500 ml of oil is to be made from sheet aluminium. There is no wastage of aluminium in producing the curved side but the circular discs for the top and bottom of the can must be cut from squares of area $(2r)^2$ where r is the radius of the can. What is the least possible area of aluminium which must be used? Answer to the nearest square centimetre.

6.9. A window is in the shape of a rectangle surmounted by a semicircle whose diameter is the width of the window. If the perimeter of the window is to be 2.5 metres, find the greatest possible area for the window to 2 decimal places.

6.10. A rectangular sheet of strengthened cardboard measures 64 cm by 40 cm. A square is to be cut from each corner of the sheet and the edges turned up to form an open box. What size of square must be cut from each corner to produce a box with maximum volume and what is that volume?

6.11. A solid sphere has a radius of 6 cm. From this sphere a regular cylinder, of height $2h$, is to be cut, see figure 6.9. Assuming that the

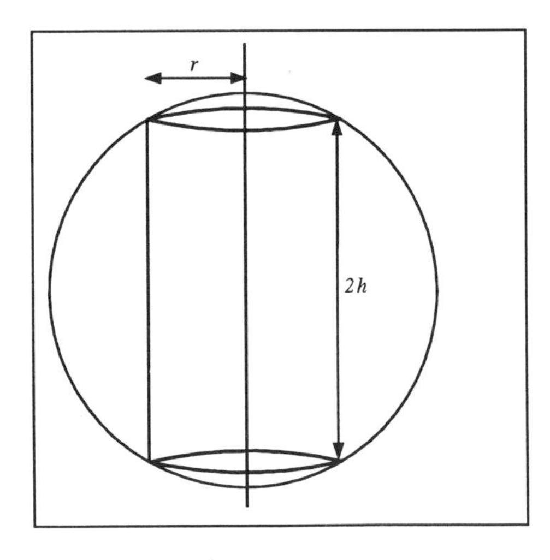

Fig. 6.9. Solid sphere

curved edges of the cylinder coincide with the surface of the sphere:
(a) find an expression for r, the radius of the cross-section of the cylinder
(b) show that the volume V of the cylinder is given by

$$V = 2\pi h(36 - h^2)$$

(c) find the maximum volume of this cylinder.

6.12. When the current through a transformer is x amps the efficiency of the transformer is given by

$$\epsilon = \frac{200x}{0.64x^2 + 200x + 350} .$$

By letting the efficiency be $\frac{1}{y}$ show that

$$y = 1 + \frac{1.75}{x} + 0.0032x .$$

Hence, find the value of x which makes y a minimum, and therefore makes the efficiency ϵ a maximum. What is the maximum efficiency? Give answers to 3 decimal place accuracy.

6.5 Taylor series

If a function $f(x)$ and its first n derivatives are continuous on an open interval containing the point $x = a$, then the nth degree *Taylor polynomial* of f about a is defined by

$$\begin{aligned} P_n(x) &= f(a) + (x-a)f'(a) + \frac{(x-a)^2}{2!}f''(a) + \ldots + \frac{(x-a)^n}{n!}f^{(n)}(a) \\ &= \sum_{j=0}^{n} \frac{(x-a)^j}{j!} f^{(j)}(a) \end{aligned}$$

and the *remainder* or difference between $P_n(x)$ and $f(x)$ is defined by

$$R_n(x) = f(x) - P_n(x) .$$

The Taylor polynomial $P_n(x)$ is an *approximation* to $f(x)$ on the open interval containing $x = a$. By including more terms (increasing n) we expect this approximation to improve, i.e. $R_n(x)$ to become smaller in magnitude.
The general form of the remainder $R_n(x)$ is given by Taylor's theorem.

Theorem 6.1

If $f(x)$ is a single–valued function such that:

(a) $f(x)$ and its first $n-1$ derivatives exist and are continuous in the closed interval $[a,b]$,
(b) $f^{(n)}(x)$ exists in the open interval (a,b)

then there exists a point $c \in (a,b)$ for which

$$\begin{aligned} f(b) &= f(a) + (b-a)f'(a) + \frac{(b-a)^2}{2!}f''(a)\ldots \\ &\quad \ldots + \frac{(b-a)^{n-1}}{(n-1)!}f^{(n-1)}(a) + \frac{(b-a)^p(b-c)^{n-p}}{p(n-1)!}f^{(n)}(c) \end{aligned}$$

where p is a fixed integer and $1 \leq p \leq n$.

Proof

The proof is along similar lines to the proof of the MVT.
Define another function $F(x)$ by

$$\begin{aligned} F(x) &= f(x) - f(b) + (b-x)f'(x) + \frac{(b-x)^2}{2!}f''(x) + \ldots \\ &\quad \ldots + \frac{(b-x)^{n-1}}{(n-1)!}f^{(n-1)}(x) + K(b-x)^p \end{aligned}$$

where K is a constant to be determined.
We have $F(b) = 0$ and we *choose* K such that $F(a) = 0$, i.e.

$$0 = f(a) - f(b) + (b-a)f'(a) + \ldots + \frac{(b-a)^{n-1}}{(n-1)!}f^{(n-1)}(a) + K(b-a)^p$$

or,

$$f(b) = f(a) + (b-a)f'(a) + \ldots + \frac{(b-a)^{n-1}}{(n-1)!}f^{(n-1)}(a) + K(b-a)^p\,. \quad (6.1)$$

Now, $F(x)$ is a function which is continuous in the closed interval $[a,b]$ and $F(a) = F(b) = 0$. Differentiating F we find that

$$F'(x) = \frac{(b-x)^{n-1}}{(n-1)!}f^{(n)}(x) - Kp(b-x)^{p-1}\,,$$

hence $F'(x)$ exists in the open interval (a,b) because $f^{(n)}(x)$ exists. Thus $F(x)$ satisfies the conditions of Rolle's theorem and so there exists a point $c \in (a,b)$ for which $F'(c) = 0$, thus

$$K = \frac{(b-c)^{n-p}}{p(n-1)!}f^{(n)}(c)\,.$$

Substituting this expression for K into equation (6.1) gives

$$\begin{aligned} f(b) &= f(a) + (b-a)f'(a) + \frac{(b-a)^2}{2!}f''(a) + \ldots \\ &\quad \ldots + \frac{(b-a)^{n-1}}{(n-1)!}f^{(n-1)}(a) + \frac{(b-a)^p(b-c)^{n-p}}{p(n-1)!}f^{(n)}(c) \end{aligned}$$

as required.□

If we consider any $x \in (a,b)$ then the continuity conditions on f and its derivatives are still satisfied on the interval (a,x) and so for any $x \in (a,b)$ we can find $c \in (a,x)$ such that

$$\begin{aligned} f(x) &= f(a) + (x-a)f'(a) + \ldots \\ &\quad \ldots + \frac{(x-a)^{n-1}}{(n-1)!}f^{(n-1)}(a) + \frac{(x-a)^p(x-c)^{n-p}}{p(n-1)!}f^{(n)}(c) \\ &= \sum_{j=0}^{n-1} \frac{(x-a)^j}{j!}f^{(j)}(a) + \frac{(x-a)^p(x-c)^{n-p}}{p(n-1)!}f^{(n)}(c) \\ &= P_{n-1}(x) + R_{n-1}(x) \end{aligned}$$

Although the remainder has been given in its general form there are two particular types which are commonly quoted.
If $p = n$ the *Lagrange* form of the remainder is obtained:

$$R_{n-1}(x) = \frac{(x-a)^n}{n!}f^{(n)}(c) .$$

If $p = 1$ the *Cauchy* form of the remainder is otained:

$$R_{n-1}(x) = \frac{(x-a)(x-c)^{n-1}}{(n-1)!}f^{(n)}(c) .$$

Problem 6.11

Find expressions for the Taylor polyomial $P_n(x)$ and the Lagrange form of the remainder $R_n(x)$ for the function $f(x) = \mathrm{e}^x$ about $x = 0$.

Solution 6.11

The function and its derivatives are identical,

$$f(x) = f'(x) = f''(x) = \ldots = f^{(n)}(x) = \mathrm{e}^x .$$

We seek the Taylor polynomial about $x = 0$, i.e. $a = 0$, so

$$f(0) = f'(0) = \ldots = f^{(n)}(0) = \mathrm{e}^0 = 1 .$$

Hence

$$\begin{aligned} P_n(x) &= f(0) + xf'(0) + \frac{x^2}{2!}f''(0) + \ldots + \frac{x^n}{n!}f^{(n)}(0) \\ &= 1 + \frac{x}{1!} + \frac{x^2}{2!} + \frac{x^3}{3!} + \ldots + \frac{x^n}{n!} . \end{aligned}$$

The Lagrange form of the remainder is

$$R_n(x) = \frac{x^{n+1}}{(n+1)!}f^{(n+1)}(c) = \frac{x^{n+1}}{(n+1)!}e^c .$$

The *Taylor series* of f about $x = a$ is the infinite series

$$f(x) = \sum_{j=0}^{\infty} \frac{(x-a)^j}{j!} f^{(j)}(a) .$$

A necessary and sufficient for the convergence of the Taylor series to $f(x)$ is that the remainder R_n tends to zero as n tends to ∞.
If $a = 0$ then we obtain the *Maclaurin series*,

$$f(x) = \sum_{j=0}^{\infty} \frac{x^j}{j!} f^{(j)}(0) .$$

The Taylor and Maclaurin series are particular examples of a power series of the form

$$c_0 + c_1(x-a) + c_2(x-a)^2 + c_3(x-a)^3 + \ldots = \sum_{j=0}^{\infty} \frac{(x-a)^j}{j!} f^{(j)}(a)$$

where the coefficients c_i are constants.

One method of determining the values of x for which a power series converges is d'Alembert's ratio test. The series is said to be absolutely convergent if

$$\lim_{n\to\infty} \left| \frac{u_{n+1}}{u_n} \right| < 1$$

where u_n and u_{n+1} are the nth and $(n+1)^{th}$ terms in the series. The convergence criteria for the general power series is then

$$\begin{aligned} \lim_{n\to\infty} \left| \frac{c_{n+1}(x-a)^{n+1}}{c_n(x-a)^n} \right| &< 1 \\ \text{or} \qquad |x-a| \lim_{n\to\infty} \left| \frac{c_{n+1}}{c_n} \right| &< 1 . \end{aligned}$$

If we write

$$\lim_{n\to\infty} \left| \frac{c_n}{c_{n+1}} \right| = R ,$$

then the convergence condition is

$$\frac{|x-a|}{R} < 1 \,,$$

and the power series is said to be absolutely convergent if $|x-a| < R$, i.e. for all $x \in (-R+a, R+a)$. The value R is called the *radius of convergence.*

Problem 6.12

Find the Maclaurin series for $f(x) = \mathrm{e}^x$.

Solution 6.12

We have already seen in Problem 6.11 that

$$\begin{aligned} f(x) &= P_n(x) + R_n(x) \\ &= 1 + \frac{x}{1!} + \frac{x^2}{2!} + \frac{x^3}{3!} + \ldots + \frac{x^n}{n!} + \frac{x^{n+1}}{(n+1)!}\mathrm{e}^c \,. \end{aligned}$$

The Maclaurin series is given by

$$f(x) = \mathrm{e}^x = 1 + \frac{x}{1!} + \frac{x^2}{2!} + \frac{x^3}{3!} + \ldots + \frac{x^n}{n!} + \ldots$$

provided the remainder tends to zero as $n \to \infty$.
Using the ratio test we have

$$\lim_{n\to\infty} \left| \frac{c_{n+1}(x-a)^{n+1}}{c_n(x-a)^n} \right| = |x| \lim_{n\to\infty} \left| \frac{c_{n+1}}{c_n} \right| = |x| \lim_{n\to\infty} \frac{1}{n+1} = 0$$

and so the remainder tends to zero as required. Notice that

$$R = \lim_{n\to\infty} \left| \frac{c_n}{c_{n+1}} \right| = \lim_{n\to\infty} \left| \frac{(n+1)!}{n!} \right| = \lim_{n\to\infty} n+1 = \infty \,.$$

The radius of convergence is infinite, i.e. the infinite series converges for all values of x. This fact is used in the following example.

Problem 6.13

Find the Taylor series expansion for $f(x) = \mathrm{e}^x$ in powers of $x - 1$.

Solution 6.13

We have already seen that

$$f(x) = f'(x) = f''(x) = \ldots = f^{(n)}(x) = \mathrm{e}^x \,.$$

We seek the Taylor series in powers of $x-1$, i.e. about $x=1$, so

$$f(1)=f'(1)=\ldots=f^{(n)}(1)=\mathrm{e}^1=\mathrm{e}\,.$$

Hence

$$\begin{aligned}
f(x) &= \mathrm{e}^x\\
&= f(1)+(x-1)f'(1)+\frac{(x-1)^2}{2!}f''(1)+\ldots+\frac{(x-1)^n}{n!}f^{(n)}(1)+\ldots\\
&= \mathrm{e}+(x-1)\mathrm{e}+\frac{(x-1)^2}{2!}\mathrm{e}+\ldots+\frac{(x-1)^n}{n!}\mathrm{e}+\ldots\\
&= \mathrm{e}\left[1+\frac{(x-1)}{1!}+\frac{(x-1)^2}{2!}+\frac{(x-1)^3}{3!}+\ldots+\frac{(x-1)^n}{n!}+\ldots\right]\\
&= \mathrm{e}\sum_{j=0}^{\infty}\frac{(x-1)^j}{j!}
\end{aligned}$$

provided the remainder tends to zero. Using the ratio test we find

$$\lim_{n\to\infty}\left|\frac{c_{n+1}(x-a)^{n+1}}{c_n(x-a)^n}\right|=|x-1|\lim_{n\to\infty}\left|\frac{c_{n+1}}{c_n}\right|=|x-1|\lim_{n\to\infty}\frac{1}{n+1}=0$$

and

$$R=\lim_{n\to\infty}\left|\frac{c_n}{c_{n+1}}\right|=\lim_{n\to\infty}\left|\frac{(n+1)!}{n!}\right|=\lim_{n\to\infty}n+1=\infty\,.$$

The radius of convergence of the Taylor series is infinite; the infinite series converges for all values of x.

EXERCISES

6.13. Expand $\sin x$ in powers of x as far as the term in x^5.
6.14. Expand $\ln(1+x)$ in powers of x as far as the term in x^5.
6.15. Expand $\mathrm{e}^{x/3}$ in powers of $x-3$ as far as the term in x^5.

7
Partial Differentiation

7.1 Introduction

In this chapter the emphasis is on functions of two independent variables but the notation and ideas extend readily to functions of three or more independent variables.

A well-known equation from physics is the gas law relating the pressure P, volume V and temperature T of an ideal gas

$$V = \frac{RT}{P}$$

where R is the universal gas constant. This equation can be rearranged in several ways, but if we consider the form above, then V is a function of T and P. If the temperature remained constant then the volume would be a function of only one independent variable P and the methods of differentiation in chapter 4 could be applied. In practice both pressure and temperature vary simultaneously and we require some method of expressing the rate of change of a function of two or more variables.

7.2 Notation

In chapter 4 we defined the notation $y = f(x)$ as 'y is a function of the single independent variable x'. Now we introduce the notation $z = f(x, y)$ meaning 'z is a function of two independent variables x and y'. In particular, for the example discussed above, $V = f(T, P)$ where $f(T, P) = RT/P$.

An advantage of using the x, y, z notation is the ability to visualise functions using three dimensional coordinate geometry. The independent variables x and y represent the position on a horizontal plane and z represents the height above (or below) this plane. This is a straightforward extension of the $y = f(x)$ approach where we interpreted $(x, f(x))$ as a curve; now $(x, y, f(x, y))$ is interpreted as a *surface*.

Consider the following examples of functions of two variables:

$$\text{(a)}\ z = x^2 + y^2 \qquad \text{(b)}\ z = x^2 - y^2 \,.$$

The first is a paraboloid, the second a hyperboloid and they are shown in figure 7.1.

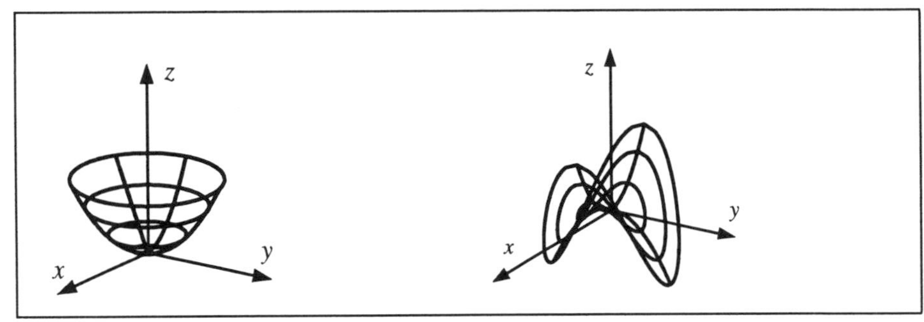

Fig. 7.1. Paraboloid and hyperboloid

In both cases z varies with respect to x and y and it seems reasonable to expect that we can find the rate of change of z with respect to both x and y.

Partial differentiation of z with respect to x means we treat y as fixed (i.e. constant) and differentiate the terms involving x as if z were a function of one variable. The formal definition of a partial derivative is very similar to the single independent variable derivative defined in chapter 4; namely,

$$\frac{\partial z}{\partial x} = \lim_{h \to 0} \frac{f(x+h, y) - f(x, y)}{h}$$

assuming the limit exists.

This is read as the partial derivative of z with respect to x. Note the difference between $\frac{\partial z}{\partial x}$ and the $\frac{dz}{dx}$ notation used in chapter 4. The ∂x indicates that z is a function of x and at least one other variable. The partial derivative with respect to y is similarly defined, assuming the limit exists, by

$$\frac{\partial z}{\partial y} = \lim_{k \to 0} \frac{f(x, y+k) - f(x, y)}{k} \,.$$

Other common notations for the partial x and y derivatives are

$$\frac{\partial f}{\partial x}, \quad f_x, \quad z_x \quad \text{and} \quad \frac{\partial f}{\partial y} hspace*5mm f_y, \quad z_y,$$

respectively. If we return to the examples above, we can write

(a) $z = x^2 + y^2$,

$\frac{\partial z}{\partial x} = 2x$, treat y as constant,

$\frac{\partial z}{\partial y} = 2y$, treat x as constant.

(b) $z = x^2 - y^2$,,

$\frac{\partial z}{\partial x} = 2x$, treat y as constant,

$\frac{\partial z}{\partial y} = -2y$, treat x as constant.

In chapters 4 and 6 we used $f'(x)$ or $\frac{dy}{dx}$ to indicate the slope of the tangent to the curve. The partial derivative notation $\frac{\partial z}{\partial x}$ indicates the slope of the surface in a direction parallel to the x-axis. Similarly, $\frac{\partial z}{\partial y}$ indicates the slope of the surface in a direction parallel to the y-axis.

Repeated partial differentiation is very similar to the one-variable case, $\frac{\partial^2 z}{\partial x^2}$, $\frac{\partial^2 z}{\partial y^2}$ being obvious extensions. The main difference is that there are more derivatives of each order, as shown in problem 7.1.

Problem 7.1

Find all first and second derivatives of the function $z = x^2 + xy$.

Solution 7.1

$\frac{\partial z}{\partial x} = 2x + y$, treat y as constant	$\frac{\partial z}{\partial y} = x$, treat x as constant,
$\frac{\partial^2 z}{\partial x^2} = 2$, treat y as constant	$\frac{\partial^2 z}{\partial y^2} = 0$, treat x as constant.

Apart from these obvious extensions of derivatives of a single variable we can also differentiate $\frac{\partial z}{\partial x}$ with respect to y to get $\frac{\partial^2 z}{\partial y \partial x}$ and differentiate $\frac{\partial z}{\partial y}$ with respect to x to get $\frac{\partial^2 z}{\partial x \partial y}$. The order of differentiation is indicated by the denominator of the derivative. In the same way that composite functions such as $fogoh$ are evaluated by applying h then g and finally f so in partial differentiation the order $\partial x \partial y$ indicates differentiation with respect to y followed by differentiation with respect to x.

In this example we have

$$\frac{\partial^2 z}{\partial x \partial y} = f_{yx} = \frac{\partial}{\partial x}\left(\frac{\partial z}{\partial y}\right) = \frac{\partial}{\partial x}(x) = 1$$

$$\text{and} \quad \frac{\partial^2 z}{\partial y \partial x} = f_{xy} = \frac{\partial}{\partial y}\left(\frac{\partial z}{\partial x}\right) = \frac{\partial}{\partial y}(2x + y) = 1 .$$

Notice that the derivatives are the same.
The following theorem is stated without proof.

Theorem 7.1

If $f(x, y)$ is a function of x and y such that f, f_x, f_y, f_{xy} and f_{yx} are continuous on an open region D, then, for every (x, y) in D,

$$\frac{\partial^2 f}{\partial x \partial y} = \frac{\partial^2 f}{\partial y \partial x} .$$

This result can save some effort, as illustrated in problem 7.2.

Problem 7.2

Find the second cross-derivative, $\frac{\partial^2 z}{\partial x \partial y}$, of the function

$$z = x^2 \cos x + x \sin y + 4 .$$

Solution 7.2

We have two options: differentiate with respect to x and then y, or vice versa. The function and all of its partial derivatives are continuous so we will end up with the same result; what we are looking for is a possible saving of effort. By inspection, there is only one term involving y and so it makes sense to differentiate with respect to y first followed by differentiation with respect to x.

$$\frac{\partial z}{\partial y} = x \cos y \quad \text{and} \quad \frac{\partial^2 z}{\partial x \partial y} = \cos y .$$

Notice that the alternative of first differentiating with respect to x involves an unnecessary differentiation of a product of functions of x.

$$\frac{\partial z}{\partial x} = 2x \cos x - x^2 \sin x + \sin y \quad \text{and} \quad \frac{\partial^2 z}{\partial y \partial x} = \cos y .$$

The order of differentiation, in general, is immaterial for continuous functions with continuous derivatives. Thus

$$\begin{aligned} f_{xxy} &= f_{yxx} = f_{xyx} \\ f_{xxyy} &= f_{yyxx} = f_{yxyx} = f_{xyyx} = \dots \end{aligned}$$

Problem 7.3

Find all first and second derivatives of $z = \cos(x - 2y)$.

Solution 7.3

$$\frac{\partial z}{\partial x} = -\sin(x-2y), \qquad \frac{\partial z}{\partial y} = 2\sin(x-2y),$$

$$\frac{\partial^2 z}{\partial x^2} = -\cos(x-2y), \qquad \frac{\partial^2 z}{\partial y^2} = -4\cos(x-2y),$$

$$\frac{\partial^2 z}{\partial x \partial y} = \frac{\partial^2 z}{\partial y \partial x} = 2\cos(x-2y)\ .$$

EXERCISES

7.1. Find the first partial derivatives, $\frac{\partial u}{\partial x}$ and $\frac{\partial u}{\partial y}$ of the following functions:
(a) $u = x^2 + y^2$
(b) $u = 2x^3 - 3x^2y + 4xy^2 - y^3$
(c) $u = \ln(3x + y)$.

7.2. Find all first and second derivatives of the function

$$z = \mathrm{e}^{2x}\sin(3y)\ .$$

7.3. If $u = x^2 - 2xy + y^2$ show that

$$\frac{\partial u}{\partial x} + \frac{\partial u}{\partial y} = 0\ .$$

7.4. A function f of x and y is said to be *harmonic* if $f_{xx} + f_{yy} = 0$. Prove that f is harmonic when

$$f(x,y) = \mathrm{e}^{-x}\cos y + \mathrm{e}^{-y}\cos x\ .$$

7.5. The volume of a cone of base radius r and perpendicular height h is given by $V = \frac{1}{3}\pi r^2 h$. Find $\frac{\partial V}{\partial r}$ and $\frac{\partial V}{\partial h}$.

7.6. The power P in watts generated in a resistance of R ohms when a voltage difference E volts is applied across the resistance is given by $P = \frac{E^2}{R}$. Find $\frac{\partial P}{\partial E}$ and $\frac{\partial P}{\partial R}$ and evaluate them when $E = 200$ volts and $R = 4000$ ohms.

7.7. The frequency f in hertz of oscillation of the current in an LC electrical circuit is given by $f = \frac{1}{2\pi\sqrt{LC}}$. Find $\frac{\partial f}{\partial L}$ and $\frac{\partial f}{\partial C}$ and evaluate them to 6 significant figure accuracy when $L = 0.25$ henrys and $C = 1.4 \times 10^{-4}$ farads.

7.3 Total differentials

The partial derivatives $\frac{\partial z}{\partial x}$ and $\frac{\partial z}{\partial y}$ represent the rates of change of z with respect to x and y. If the independent variables, x and y, are changed by increments Δx and Δy the differentials dx and dy are defined by $dx = \Delta x$, $dy = \Delta y$.
If x varies and y is held fixed then z is a function of x only and the *partial differential* of z with respect to x is defined by

$$\frac{\partial z}{\partial x}dx \; .$$

Similarly, the partial differential of z with respect to y is defined by

$$\frac{\partial z}{\partial y}dy \; .$$

The *total differential* dz is then defined by

$$dz = \frac{\partial z}{\partial x}dx + \frac{\partial z}{\partial y}dy \; .$$

When the increments of the independent variables are small, the total differential gives a good approximation to the increment Δz of the function.

Problem 7.4

If $z = xy$ find the total differential dz if x is changed from 5 to 5.1 and y from 2 to 2.01 .

Solution 7.4

We have $dx = \Delta x = 0.1$, $dy = \Delta y = 0.01$. The first derivatives of z are $\frac{\partial z}{\partial x} = y$ and $\frac{\partial z}{\partial y} = x$.
Substituting these into the expression for the total differential gives

$$dz = \frac{\partial z}{\partial x}dx + \frac{\partial z}{\partial y}dy = ydx + xdy \; .$$

Substituting $x = 5$, $y = 2$, $dx = \Delta x = 0.1$ and $dy = \Delta y = 0.01$ gives

$$dz = 2 \times 0.1 + 5 \times 0.01 = 0.25 \; .$$

Increasing x by 0.1 and y by 0.01 has resulted in an increase of z of *approximately* 0.25 . The *actual* change can be calculated directly:

$$\begin{aligned} x = 5 \text{ and } y = 2 \quad & \text{gives } z = xy = 10 \\ x = 5.1 \text{ and } y = 2.01 \quad & \text{gives } z = xy = 10.251 \; . \end{aligned}$$

So the increment in z was 0.251.

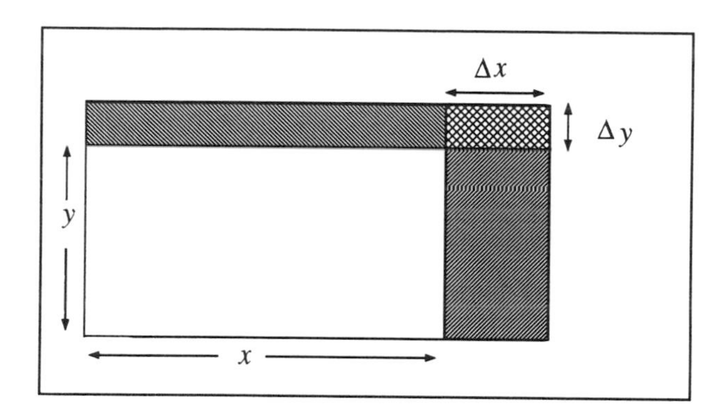

Fig. 7.2. Illustration of a total differential

A geometric interpretation of this example is illustrated in figure 7.2.

Let x and y be the lengths of the sides of a rectangle, then $z = xy$ is the area of the rectangle. The terms

$$\frac{\partial z}{\partial x}dx = ydx \quad \text{and} \quad \frac{\partial z}{\partial y}dy = xdy$$

are the areas of the two larger rectangles that have been added. By calculating the total differential and using it to find the approximate change in area we have neglected the small rectangle at the top right corner whose area is $dxdy$.

This approach relies on *small* changes so that the terms $\frac{\partial z}{\partial x}dx$ and $\frac{\partial z}{\partial y}dy$ are small compared to the original function value.

Problem 7.5

The power consumed in a resistor is $P = \frac{E^2}{R}$ watts. If $E = 200$ volts and $R = 8$ ohms by how much does the power consumption change if E is decreased by 5 volts and R is decreased by 0.2 ohms ?

Solution 7.5

Looking at the equation for P we see that decreasing E will reduce P but decreasing R will increase P. Whether P increases or decreases will depend on the relative effects of E and R.

The values given are

$$E = 200, \quad dE = -5, \quad R = 8, \quad dR = -0.2\,.$$

The partial derivatives are

$$\frac{\partial P}{\partial E} = \frac{2E}{R}, \qquad \frac{\partial P}{\partial R} = -\frac{E^2}{R^2}\,.$$

Therefore the total differential is

$$\begin{aligned} dP &= \frac{\partial P}{\partial E}dE + \frac{\partial P}{\partial R}dR = \frac{2E}{R}dE - \frac{E^2}{R^2}dR \\ &= \frac{2 \times 200}{8}(-5) - \frac{200^2}{8^2}(-0.2) = -250 + 125 = -125 \text{ watts} \end{aligned}$$

where the negative value indicates a decrease in power consumption.

Notice the final value of -125 was made of two parts, -250 due to the decrease in E and $+125$ due to the decrease in R. This partial cancelling effect is common in functions of two or more variables.

EXERCISES

7.8. If x is increased from 1 to 1.01 and y is decreased from 2 to 1.98, find the approximate change in z where $z(x, y)$ is given by

$$z = 2x^2 - 3xy + 5y^2 - 4 .$$

Check your result by calculating $z(1,2)$ and $z(1.01, 1.98)$.

7.9. The fundamental frequency of vibration n of a wire of circular cross-section under a tension T is

$$n = \frac{1}{2rl}\sqrt{\frac{T}{\pi d}}$$

where l is the length, r the radius and d the density of the wire. Find the approximate effect on n of changing l by a small amount dl and T by a small amount dT.

7.4 The total derivative

This is the chain rule for functions of functions. It is a natural extension of the definition of the total differential. Instead of specified differentials dx, dy leading to a total differential dz, the variables x and y are now functions of another variable t so that we have *rates of change* $\frac{dx}{dt}$, $\frac{dy}{dt}$ leading to a *rate of change* $\frac{dz}{dt}$.

$$\frac{dz}{dt} = \frac{\partial z}{\partial x}\frac{dx}{dt} + \frac{\partial z}{\partial y}\frac{dy}{dt}$$

Problem 7.6

The height of a right circular cone is 15 cm and is increasing at 0.2 cm/s. The radius of the base is 10 cm and is decreasing at 0.3 cm/s. At what rate is the volume changing?

Solution 7.6

Let r represent the radius of the base and h the height of the cone. The volume V is given by

$$V = \frac{1}{3}\pi r^2 h\,.$$

It should be obvious from the formula that decreasing r decreases V and conversely increasing h increases V.

Whether the changes in r and h result in an overall volume increase or decrease depends on the relative effects of r and h. We are trying to find the rate of change of V with respect to time using

$$\frac{dV}{dt} = \frac{\partial V}{\partial r}\frac{dr}{dt} + \frac{\partial V}{\partial h}\frac{dh}{dt}\,.$$

We first find the two partial derivatives:

$$\frac{\partial V}{\partial r} = \frac{2}{3}\pi r h, \qquad \frac{\partial V}{\partial h} = \frac{1}{3}\pi r^2\,.$$

The rates of change of r and h with time are $\frac{dr}{dt} = -0.3$ and $\frac{dh}{dt} = 0.2$. Hence using the values $r = 10$ cm and $h = 15$ cm we get

$$\begin{aligned}\frac{dV}{dt} &= \frac{\partial V}{\partial r}\frac{dr}{dt} + \frac{\partial V}{\partial h}\frac{dh}{dt} = \left(\frac{2}{3}\pi r h\right)(-0.3) + \left(\frac{1}{3}\pi r^2\right)(0.2)\\ &= -\frac{90\pi}{3} + \frac{20\pi}{3} = -\frac{70\pi}{3}\ \mathrm{cm}^3/\mathrm{s}\,.\end{aligned}$$

EXERCISES

7.10. Find $\frac{dz}{dt}$ given $x = \sin t$, $y = \cos t$ and $z = 2x^2 + 3xy - 4y^2 + 15$.

7.11. The equation $\epsilon = \alpha(BR)^{1/2}$ arises in the study of thermal noise. If B and R are functions of time t, write down an expression, in its simplest form, for $\frac{d\epsilon}{dt}$.

7.5 Taylor series for functions of two variables

The Taylor series expansion for a function of a single variable, about the point $x = a$, is

$$f(x) = f(a) + (x-a)f'(a) + \frac{(x-a)^2}{2!}f''(a) + \ldots + \frac{(x-a)^n}{n!}f^{(n)}(a) + \ldots$$

A particular example is the exponential function which has the Taylor series

$$e^x = 1 + \frac{x}{1!} + \frac{x^2}{2!} + \frac{x^3}{3!} + \ldots + \frac{x^n}{n!} + \ldots$$

when expanded about the point $x = 0$.

For functions of two variables, the expressions appear more complex but the ideas are the same – there are simply more terms because of the greater number of derivatives. We will not attempt to derive the general expression but will simply state the first few terms. The expansion about the point (a, b) is

$$\begin{aligned} f(x,y) = f(a,b) &+ [(x-a)f_x(a,b) + (y-b)f_y(a,b)] \\ &+ \frac{1}{2!}[(x-a)^2 f_{xx}(a,b) + 2(x-a)(y-b)f_{xy}(a,b) \\ &\qquad +(y-b)^2 f_{yy}(a,b)] \\ &+ \frac{1}{3!}[(x-a)^3 f_{xxx}(a,b) + 3(x-a)^2(y-b)f_{xxy}(a,b) \\ &\qquad +3(x-a)(y-b)^2 f_{xyy}(a,b) + (y-b)^3 f_{yyy}(a,b)] \\ &+ \ldots \end{aligned}$$

Note that if $x = a$ or $y = b$ then the above reduces to the Taylor series for a single independent variable.

As for the Taylor series expansion in chapter 6 there are several forms of the series. The most common of the other variants is to expand about a general point (x, y) for small increments h and k in x and y respectively.

$$\begin{aligned} f(x+h, y+k) = f(x,y) &+ [hf_x(x,y) + kf_y(x,y)] \\ &+ \frac{1}{2!}\left[h^2 f_{xx}(x,y) + 2hkf_{xy}(x,y) + k^2 f_{yy}(x,y)\right] \\ &+ \frac{1}{3!}[h^3 f_{xxx}(x,y) + 3h^2 k f_{xxy}(x,y) \\ &\qquad +3hk^2 f_{xyy}(x,y) + k^3 f_{yyy}(x,y)] \\ &+ \ldots \end{aligned}$$

Problem 7.7

Express the polynomial $f(x,y) = x^3 + x^2y - 3y^2x + 5xy - 10$ in terms of its derivatives at the point $x = 0, y = 0$. By evaluating these derivatives at $x = 0, y = 0$ show that it reduces to the given polynomial.

Solution 7.7

Find the partial derivatives up to the third order.

$$\begin{aligned}
f_x &= 3x^2 + 2xy - 3y^2 + 5y, & f_y &= x^2 - 6xy + 5x,\\
f_{xx} &= 6x + 2y, & f_{yy} &= -6x,\\
f_{xy} &= 2x - 6y + 5, & &\\
f_{xxx} &= 6, & f_{yyy} &= 0,\\
f_{xxy} &= 2, & f_{xyy} &= -6.
\end{aligned}$$

All higher derivatives are identically zero.
The Taylor series expansion about the point $(0,0)$ is

$$\begin{aligned}
f(x,y) = f(0,0) \quad &+ \quad [x f_x(0,0) + y f_y(0,0)]\\
&+ \quad \frac{1}{2!}[x^2 f_{xx}(0,0) + 2xy f_{xy}(0,0) + y^2 f_{yy}(0,0)]\\
&+ \quad \frac{1}{3!}[x^3 f_{xxx}(0,0) + 3x^2 y f_{xxy}(0,0)\\
&\qquad\qquad + 3xy^2 f_{xyy}(0,0) + y^3 f_{yyy}(0,0)]
\end{aligned}$$

For a polynomial this should be exact and all we are doing is rewriting it in terms of its derivatives. Evaluating the derivatives at $(0,0)$:

$f_x = 0, \quad f_y = 0,$
$f_{xx} = 0, \quad f_{yy} = 0, \quad f_{xy} = 5,$
$f_{xxx} = 6, \quad f_{yyy} = 0, \quad f_{xxy} = 2, \quad f_{xyy} = -6.$

Hence

$$\begin{aligned}
f(x,y) &= -10 + \frac{10xy}{2!} + \frac{6x^3 + 6x^2y - 18xy^2}{3!}\\
&= -10 + 5xy + x^3 + x^2y - 3xy^2\\
&= x^3 + x^2y - 3y^2x + 5xy - 10\ .
\end{aligned}$$

as required.

Problem 7.8

Find the Taylor series expansion, in powers of x and y, as far as the third derivatives for the function

$$f(x,y) = e^{x+y}\ .$$

Solution 7.8

The function and all its partial derivatives are identical.

$$\begin{aligned} f_x &= f_y = e^{x+y} \\ f_{xx} &= f_{yy} = f_{xy} = e^{x+y} \\ f_{xxx} &= f_{yyy} = f_{xxy} = f_{xyy} = e^{x+y} \end{aligned}$$

At $x = y = 0$ the exponential e^{x+y} is unity and hence the expansion about the point $(0,0)$ is

$$\begin{aligned} f(x,y) &= 1 + (x+y) + \frac{x^2 + 2xy + y^2}{2!} + \frac{x^3 + 3x^2y + 3y^2x + y^3}{3!} + \dots \\ &= 1 + (x+y) + \frac{(x+y)^2}{2!} + \frac{(x+y)^3}{3!} + \dots \end{aligned}$$

The equivalent expansion for increments h and k about the point (x,y) is

$$f(x+h, y+k) = e^{x+y}\left[1 + (h+k) + \frac{(h+k)^2}{2!} + \frac{(h+k)^3}{3!} + \dots\right]$$

which is very similar to the expansion for a single independent variable case.

Problem 7.9

Find the Taylor series expansion in powers of x and y as far as the third derivatives for the function

$$z = \sin(2x + y)\ .$$

Solution 7.9

First find the partial derivatives:

$$\begin{aligned} &z_x = 2\cos(2x+y), && z_y = \cos(2x+y), \\ &z_{xx} = -4\sin(2x+y), && z_{yy} = -\sin(2x+y), \\ &z_{xy} = -2\sin(2x+y), && \\ &z_{xxx} = -8\cos(2x+y), && z_{yyy} = -\cos(2x+y), \\ &z_{xxy} = -4\cos(2x+y), && z_{xyy} = -2\cos(2x+y). \end{aligned}$$

The expansion is then

$$\begin{aligned} z(x,y) &= z(0,0) + [xz_x(0,0) + yz_y(0,0)] \\ &\quad + \frac{1}{2!}\left[x^2 z_{xx}(0,0) + 2xyz_{xy}(0,0) + y^2 z_{yy}(0,0)\right] + \dots \end{aligned}$$

At $x = y = 0$ we have $\sin(2x + y) = 0$ and $\cos(2x + y) = 1$ leading to

$$\begin{aligned} z(x,y) &= \frac{2x+y}{1!} - \frac{8x^3 + 12x^2y + 6xy^2 + y^3}{3!} + \ldots \\ &= \frac{2x+y}{1!} - \frac{(2x)^3 + 3(2x)^2y + 3(2x)y^2 + y^3}{3!} + \ldots \end{aligned}$$

Again, if x or y were constant the above expression would reduce to the Taylor expansion of a single variable.

EXERCISES

7.12. Find the Taylor series expansion in powers of x and y as far as the third derivatives for the function

$$f(x,y) = \ln(1 + x + y) \; .$$

7.13. Find the Taylor series expansion in powers of x and y as far as the third derivatives for the function

$$z = \mathrm{e}^{x^2+y} \; .$$

7.6 Maxima, minima and saddlepoints

In chapter 6 we analysed functions of a single variable and found maxima, minima and points of inflection. In this section we extend these ideas to functions of two variables.

Let f be a function defined on a region containing the point (x_0, y_0). Then

1. $f(x_0, y_0)$ is a *relative maximum* of f if $f(x_0, y_0) \geq f(x, y)$ for all (x, y) in an open disc containing (x_0, y_0).
2. $f(x_0, y_0)$ is a *relative minimum* of f if $f(x_0, y_0) \leq f(x, y)$ for all (x, y) in an open disc containing (x_0, y_0).

The difference between the single-variable function and the two-variable function, as far as the definitions of relative maximum and relative minimum are concerned, is that an *open interval* becomes an *open disc* due to the increase in dimensionality.

7.6.1 Classification of critical points

Let $f(x, y)$ have continuous first and second derivatives on an open region containing a point (x_0, y_0) for which $f_x(x_0, y_0) = f_y(x_0, y_0) = 0$. Define d by

$$d = [f_{xy}(x_0, y_0)]^2 - f_{xx}(x_0, y_0) \times f_{yy}(x_0, y_0) \ .$$

1. If $d < 0$ and $f_{xx}(x_0, y_0) > 0$ then there is a *minimum* at (x_0, y_0).
2. If $d < 0$ and $f_{xx}(x_0, y_0) < 0$ then there is a *maximum* at (x_0, y_0).
3. If $d > 0$ there is a *saddlepoint* at (x_0, y_0).
4. If $d = 0$ the test fails.

A derivation of this result is left as exercise 7.16.

Method

The suggested method for finding critical points and then classifying them is

1. Find first derivatives.
2. Find second derivatives.
3. Solve $f_x = f_y = 0$ to find any critical points.
4. Use the second derivatives to determine their nature.

Problem 7.10

Find the location and nature of the critical points of the function

$$f(x, y) = x^3 + 3y^2x - 3x^2 - 3y^2 + 4 \ .$$

Solution 7.10

The first derivatives of the function are

$$\frac{\partial f}{\partial x} = f_x = 3x^2 + 3y^2 - 6x, \qquad \frac{\partial f}{\partial y} = f_y = 6xy - 6y \ .$$

To determine the nature of these critical points we need the second partial derivatives of the function

$$\frac{\partial^2 f}{\partial x^2} = f_{xx} = 6x - 6 \ , \ \frac{\partial^2 f}{\partial y^2} = f_{yy} = 6x - 6 \ , \ \frac{\partial^2 f}{\partial x \partial y} = f_{xy} = 6y \ .$$

Both first derivatives must be zero for critical points, hence

$$\begin{aligned} 3(x^2 + y^2 - 2x) &= 0 \\ 6y(x - 1) &= 0 \end{aligned}$$

From the second of these equations we see that $y = 0$ or $x = 1$.
Substituting $y = 0$ into the other equation gives $x^2 - 2x = 0$ and so $x = 0$ or $x = 2$.
Substituting $x = 1$ gives $y^2 - 1 = 0$ and so $y = \pm 1$.
Hence the critical points of f are :

$$(0,0), \quad (1,1), \quad (1,-1), \quad (2,0)$$

Create an appropriate table:

	(0,0)	(2,0)	(1,1)	(1, −1)
$f_{xx} = 6x - 6$	−6	6	0	0
$f_{xy} = 6y$	0	0	6	−6
$f_{yy} = 6x - 6$	−6	6	0	0
$d = (f_{xy})^2 - f_{xx}f_{yy}$	<0	<0	>0	>0
	max	min	saddle	saddle

Problem 7.11

Find and characterise all critical points of the following function

$$f(x, y) = \sin x \sin y + \cos^2 x \,, \qquad 0 < x < \pi, \quad 0 < y < 2\pi \,.$$

Solution 7.11

The first step is to find the first derivatives.

$$f_x = \sin y \cos x - 2 \sin x \cos x \,, \qquad f_y = \sin x \cos y \,.$$

We need the second derivatives to classify the critical points.

$$\begin{aligned} f_{xx} &= 2\sin^2 x - 2\cos^2 x - \sin y \sin x \\ f_{yy} &= -\sin x \sin y \\ f_{xy} &= \cos x \cos y \end{aligned}$$

The first derivatives are zero at a critical point so we are looking for solutions to the pair of equations

$$\begin{aligned} \cos x(\sin y - 2\sin x) &= 0 \\ \sin x \cos y &= 0 \end{aligned}$$

In the second equation there are no values of x in the *open interval* $(0, \pi)$ for which $\sin x = 0$, but $\cos y = 0$ when $y = \pi/2$ and $y = 3\pi/2$.

Substituting these values into the first equation we get

$$y = \tfrac{\pi}{2}: \quad \cos x(1 - 2\sin x) = 0, \qquad \cos x = 0 \text{ or } \sin x = \tfrac{1}{2},$$

$$y = \tfrac{3\pi}{2}: \quad -\cos x(1 + 2\sin x) = 0, \quad \cos x = 0 \text{ or } \sin x = -\tfrac{1}{2},$$

and the x-coordinates of the critical points are:

when $y = \frac{\pi}{2}$ $\quad x = \frac{\pi}{2}, \frac{\pi}{6}, \frac{5\pi}{6}$

when $y = \frac{3\pi}{2}$ $\quad x = \frac{\pi}{2}$ only

Create an appropriate table:

	$(\frac{\pi}{2}, \frac{\pi}{2})$	$(\frac{\pi}{6}, \frac{\pi}{2})$	$(\frac{5\pi}{6}, \frac{\pi}{2})$	$(\frac{\pi}{2}, \frac{3\pi}{2})$
$f_{xx} = 2\sin^2 x - 2\cos^2 x$ $- \sin y \sin x$	1	$-\frac{3}{2}$	$-\frac{3}{2}$	3
$f_{yy} = -\sin x \sin y$	-1	$-\frac{1}{2}$	$-\frac{1}{2}$	1
$f_{xy} = \cos x \cos y$	0	0	0	0
$d = (f_{xy})^2 - f_{xx}f_{yy}$	>0	<0	<0	<0
	saddle	max	max	min

EXERCISES

7.14. Find and classify the critical points of

$$f(x, y) = x^2 - 8x + y^2 + 9\,.$$

7.15. Find and classify the critical points of

$$f(x, y) = 4xy - x^4 - y^4\,.$$

7.16. Derive the second derivative test:

(a) If $d < 0$ and $f_{xx}(x_0, y_0) > 0$ then there is a *minimum* at (x_0, y_0).

(b) If $d < 0$ and $f_{xx}(x_0, y_0) < 0$ then there is a *maximum* at (x_0, y_0).

(c) If $d > 0$ there is a *saddlepoint* at (x_0, y_0).

[Hint: assume the expansion

$$\begin{aligned} f(x+h, y+k) &\approx f(x,y) + [hf_x(x,y) + kf_y(x,y)] \\ &+ \frac{1}{2}\left[h^2 f_{xx}(x,y) + 2hkf_{xy}(x,y) + k^2 f_{yy}(x,y)\right] \end{aligned}$$

is sufficiently accurate for small h, k and then consider what determines the *sign* of $f(x+h, y+k) - f(x, y)$ at a critical point (x_0, y_0). Multiply both sides by $f_{xx}(x_0, y_0)$ and complete the square.]

7.7 Problems with constraints

Minimisation or maximisation problems are usually subject to constraints. So far we have only developed tests for classifying extrema of functions of two variables. If a function has more than two variables we usually consider one of two options. The first is to use the constraint to reduce the function to have only two independent variables and the second is to use the method of Lagrange multipliers.

In this section we will use the first method which is a straightforward extension of the ideas used in section 6.5 for functions of one independent variable.

Problem 7.12

Find three positive numbers x, y and z such that $x + y + z = 8$ and the function $W = x^2yz$ is maximum.

Solution 7.12

W is a function of three variables but we can use the constraint to eliminate one of the variables and then find the relative extrema of the function of two independent variables. We can choose to eliminate any of the variables because all three appear in the constraint. Suppose we decide to eliminate z, then we have

$$\begin{aligned} z &= 8 - x - y, \\ W &= x^2yz = x^2y(8 - x - y) \\ &= 8x^2y - x^3y - x^2y^2 . \end{aligned}$$

We need the first derivatives to find the critical point

$$\begin{aligned} \frac{\partial W}{\partial x} &= 16xy - 3x^2y - 2xy^2 = xy(16 - 3x - 2y), \\ \frac{\partial W}{\partial y} &= 8x^2 - x^3 - 2x^2y = x^2(8 - x - 2y) . \end{aligned}$$

The second derivatives will be needed to classify the critical points.

$$\begin{aligned} \frac{\partial^2 W}{\partial x^2} &= 16y - 6xy - 2y^2, \\ \frac{\partial^2 W}{\partial y^2} &= -2x^2, \\ \frac{\partial^2 W}{\partial x \partial y} &= 16x - 3x^2 - 4xy . \end{aligned}$$

The critical points occur where the first derivatives are zero

$$\begin{aligned}\frac{\partial W}{\partial x} &= xy(16 - 3x - 2y) = 0, \\ \frac{\partial W}{\partial y} &= x^2(8 - x - 2y) = 0\ .\end{aligned}$$

As we are trying to maximise W we cannot have zero values for x and y, so the values of x and y are given by the solution of the pair of simultaneous equations

$$\begin{aligned} 16 - 3x - 2y &= 0 \\ 8 - x - 2y &= 0\ . \\ \text{Subtract :}\qquad 8 - 2x &= 0\ . \end{aligned}$$

Hence $x = 4$ and $y = 2$. The value of z is obtained by substituting these values into the constraint, which gives $z = 2$.

Although we have found a critical point we must check that it is a relative maximum by using the second derivative test. Evaluating the second derivatives at $x = 4$, $y = 2$ gives:

$$\begin{aligned}\frac{\partial^2 W}{\partial x^2} &= 16y - 6xy - 2y^2 = 32 - 48 - 8 = -24, \\ \frac{\partial^2 W}{\partial y^2} &= -2x^2 = -32, \\ \frac{\partial^2 W}{\partial x \partial y} &= 16x - 3x^2 - 4xy = 64 - 48 - 32 = -16,\end{aligned}$$

and hence

$$d = \left[\frac{\partial^2 W}{\partial x \partial y}\right]^2 - \frac{\partial^2 W}{\partial x^2}\frac{\partial^2 W}{\partial y^2} = (-16)^2 - (-24)(-32) = -512$$

The negative values for both d and f_{xx} indicates a relative maximum and so the solution is $x = 4$, $y = 2$, $z = 2$.

Problem 7.13

The square of the distance, s, of a point (x, y, z) from the origin is

$$s^2 = x^2 + y^2 + z^2\ .$$

By minimising s^2 subject to the restriction $x + 2y - z = 10$, find the point on the plane $x + 2y - z = 10$ which is nearest to the origin.

Solution 7.13

As in the previous problem we use the constraint to eliminate one of the independent variables from the function. Again, eliminate z.

$$\begin{aligned} z &= 10 - x - 2y, \\ s^2 &= x^2 + y^2 + z^2 = x^2 + y^2 + (10 - x - 2y)^2 \\ &= 2x^2 + 5y^2 - 20x - 40y + 4xy + 100 . \end{aligned}$$

We now find the first and second partial derivatives of s^2.

$$\frac{\partial s^2}{\partial x} = 4x + 4y - 20, \quad \frac{\partial s^2}{\partial y} = 10y + 4x - 40,$$

$$\frac{\partial^2 s^2}{\partial x^2} = 4, \qquad \frac{\partial^2 s^2}{\partial y^2} = 10, \qquad \frac{\partial^2 s^2}{\partial x \partial y} = 4 .$$

All the second derivatives are constants and so we can determine the nature of any critical points before we find them.

$$d = \left[\frac{\partial^2 s^2}{\partial x \partial y}\right]^2 - \frac{\partial^2 s^2}{\partial x^2}\frac{\partial^2 s^2}{\partial y^2} = (4)^2 - (4)(10) = -24 .$$

The negative value of d and positive value of $\frac{\partial^2 s^2}{\partial x^2}$ indicates any critical points are relative minima.
The critical points occur where the first derivatives are zero, that is

$$\begin{aligned} \frac{\partial s^2}{\partial x} &= 4x + 4y - 20 = 0, \\ \frac{\partial s^2}{\partial y} &= 10y + 4x - 40 = 0 . \\ \text{Subtract :} \qquad -6y + 20 &= 0 . \end{aligned}$$

Hence $y = \frac{10}{3}$ and $x = \frac{5}{3}$. The value of z is obtained by substituting these values into the constraint, which gives $z = -\frac{5}{3}$.
We have already ascertained that any critical points are relative minima, therefore $(\frac{5}{3}, \frac{10}{3}, -\frac{5}{3})$ is the point on the plane closest to the origin.

EXERCISES

7.17. Find the values of x, y and z for which the function

$$f(x, y, z) = x^2 + y^2 + z^2$$

has a minimum value subject to the constraint $x + 3y - 2z = 4$.

7.18. For a package to go by parcel post, Post Office regulations stipulate that the length plus the circumference must not exceed 2 m. (The circumference is the perimeter of the cross-section.) Find the largest volume which can be sent by parcel post in a rectangular box.

7.8 Lagrange multipliers

Another method of taking constraints into account is Lagrange's method of undetermined multipliers, usually abbreviated to Lagrange multipliers. This method has the advantage of being more general and can find the extrema of a function of an arbitrary number of variables subject to a list of constraints.

The general problem can be posed in the form:

Find the extrema of the function $f(x_1, x_2, \ldots, x_n)$ of n variables subject to m constraints $c_r(x_1, x_2, \ldots, x_n) = 0$, where $r = 1, 2, \ldots, m$.

A necessary condition for local extrema of a function $F(x_1, x_2, \ldots, x_n)$, when $x_1, \ldots, x_n$ are independent variables, is

$$\frac{\partial F}{\partial x_1} = \frac{\partial F}{\partial x_2} = \ldots = \frac{\partial F}{\partial x_n} = 0 \, ,$$

which is merely an extension of the theory for a function of two variables.

We cannot apply this to our function $f(x_1, x_2, \ldots, x_n)$ because the variables $x_1, x_2, \ldots, x_n$ are *not* independent; they are related via the constraints c_r.

To solve this problem we introduce m multipliers $\lambda_1, \lambda_2, \ldots \lambda_m$ and try to minimise a new function g defined by

$$g(x_1, x_2, \ldots, x_n, \lambda_1, \ldots, \lambda_m) = f(x_1, x_2, \ldots, x_n) + \sum_{r=1}^{m} \lambda_r c_r(x_1, x_2, \ldots, x_n) \, .$$

The new variables λ_r are called the Lagrange multipliers and g is a function of $n + m$ variables. The critical points of this alternative problem are found by solving the $n + m$ equations

$$\frac{\partial g}{\partial x_i} = \frac{\partial f}{\partial x_i} + \sum_{r=1}^{m} \lambda_r \frac{\partial c_r}{\partial x_i} = 0, \qquad i = 1, 2, \ldots, n \, , \tag{7.1}$$

$$\frac{\partial g}{\partial \lambda_r} = c_r = 0, \qquad r = 1, 2, \ldots, m \tag{7.2}$$

and then determining which of these give solutions to the original problem. Notice that $\dfrac{\partial g}{\partial \lambda_r} = c_r = 0$ forces the constraints to be satisfied.

By constructing the function g in this way, the critical points are now influenced by the function f *and* the constraints c_r. Notice also that if one of the variables, say x_1, did not appear in any of the constraints then x_1 would be a truly independent variable and so

$$\frac{\partial g}{\partial x_1} = \frac{\partial f}{\partial x_1} .$$

In practice it is extremely difficult to classify these critical points except for relatively simple problems such as those considered in the previous section. To illustrate the method we will redo problem 7.12 using the above notation.

Problem 7.14

Find the critical points of the function

$$f(x_1, x_2, x_3) = x_1^2 \, x_2 \, x_3$$

subject to the constraint $x_1 + x_2 + x_3 = 8$.

Solution 7.14

We assume that x_1, x_2 and x_3 are non-zero, otherwise the function is identically zero and any optimisation process is meaningless. There is only one constraint so we are seeking one multiplier, λ_1. The problem now is to find the critical points of the function defined by

$$g = x_1^2 x_2 x_3 + \lambda_1 (x_1 + x_2 + x_3 - 8) .$$

There are four unknowns, x_1, x_2, x_3 and λ_1 so we find the four partial derivatives and set them equal to zero to find the critical points.

$$\begin{aligned}
\frac{\partial g}{\partial x_1} &= 2x_1 x_2 x_3 + \lambda_1 = 0 \\
\frac{\partial g}{\partial x_2} &= x_1^2 x_3 + \lambda_1 = 0 \\
\frac{\partial g}{\partial x_3} &= x_1^2 x_2 + \lambda_1 = 0 \\
\frac{\partial g}{\partial \lambda_1} &= x_1 + x_2 + x_3 - 8 = 0
\end{aligned}$$

We have a set of four simultaneous equations to solve. The middle pair give us a link between x_2 and x_3

$$\begin{aligned}
x_1^2 x_3 + \lambda_1 &= 0 \\
x_1^2 x_2 + \lambda_1 &= 0 \\
\text{Subtract} \quad x_1^2 (x_3 - x_2) &= 0
\end{aligned}$$

and because $x_1 \neq 0$ we must have $x_2 = x_3$. The simultaneous equations are now

$$\begin{aligned} 2x_1x_2^2 + \lambda_1 &= 0 \\ x_1^2x_2 + \lambda_1 &= 0 \\ x_1 + 2x_2 - 8 &= 0\,. \end{aligned}$$

Taking the first pair we find the link between x_1 and x_2

$$\begin{aligned} 2x_1x_2^2 + \lambda_1 &= 0 \\ x_1^2x_2 + \lambda_1 &= 0 \\ \text{Subtract} \quad x_1x_2(2x_2 - x_1) &= 0 \end{aligned}$$

By the same reasoning as before we must have $2x_2 = x_1$ and so the constraint $x_1 + 2x_2 - 8 = 0$ now becomes

$$2x_1 - 8 = 0 \Rightarrow x_1 = 4\,.$$

Back substitution gives us $x_2 = 2$, $x_3 = 2$ and $\lambda_1 = -32$. Notice that in this example we never required the value of λ_1.

Problem 7.15

Find the critical points of the function

$$f(x,y,z) = x^2 + y^2 + z^2$$

subject to the constraints $x + y + z = 1$ and $xyz = -1$.

Solution 7.15

There are two constraints so we are seeking two multipliers, λ_1 and λ_2. We are trying to find the critical points of the function defined by

$$g = x^2 + y^2 + z^2 + \lambda_1(x + y + z - 1) + \lambda_2(xyz + 1)\,.$$

There are five unknowns, x_1, x_2, x_3, λ_1 and λ_2 so we find the partial derivatives and set them equal to zero to find the critical points.

$$\begin{aligned} \frac{\partial g}{\partial x} &= 2x + \lambda_1 + \lambda_2 yz = 0 \\ \frac{\partial g}{\partial y} &= 2y + \lambda_1 + \lambda_2 xz = 0 \\ \frac{\partial g}{\partial z} &= 2z + \lambda_1 + \lambda_2 xy = 0 \\ \frac{\partial g}{\partial \lambda_1} &= x + y + z - 1 = 0 \\ \frac{\partial g}{\partial \lambda_2} &= xyz + 1 = 0\,. \end{aligned}$$

Eliminating λ_1 and λ_2 from the first three equations gives

$$(x - y)(y - z)(z - x) = 0$$

so we must have $x = y$ or $y = z$ or $x = z$.

When $x = y$, the constraints become

$$2y + z = 1 \quad \text{and} \quad y^2 z = -1$$

and if we eliminate z we obtain

$$2y^3 - y^2 - 1 = (y - 1)(2y^2 + y + 1) = 0$$

which has only one real solution, $y = 1$. Thus $x = y = 1$, $z = -1$ is a critical point.

When $y = z$, the constraints become

$$x + 2z = 1 \quad \text{and} \quad xz^2 = -1$$

and if we eliminate x we obtain

$$2z^3 - z^2 - 1 = (z - 1)(2z^2 + z + 1) = 0$$

which has only one real solution, $z = 1$. Thus $x = -1$, $y = z = -1$ is a critical point.

Similarly, when $z = x$, the constraints become

$$2x + y = 1 \quad \text{and} \quad x^2 y = -1$$

and if we eliminate y we obtain

$$(x - 1)(2x^2 + x + 1) = 0$$

which has only one real solution, $x = 1$. Thus $x = z = 1$, $y = -1$ is a critical point.

There are three critical points of g, all of which yield the same value of the function f.

EXERCISES

7.19. Find the critical points of the function

$$f(x_1, x_2, x_3) = 4x_1x_2 + 6x_1x_3 + 4x_2x_3$$

subject to the constraint $x_1x_2x_3 = 12$.

7.20. Find the critical points of the function

$$f(x, y, z) = xy^2z^2$$

subject to the constraints $x + y + z = 6$, $x > 0$, $y > 0$, $z > 0$.

8
Multiple Integrals

8.1 Double integrals

Consider a function $z = f(x, y)$ which is defined and continuous on the rectangular region $a \leq x \leq b$, $c \leq y \leq d$. We will simplify the notation further by assuming that z only takes positive values on this region R, i.e. we have a surface $(x, y, f(x, y))$ above the horizontal x, y plane. We wish to calculate the volume enclosed between the rectangular region of the x, y plane and this surface, as illustrated in figure 8.1.

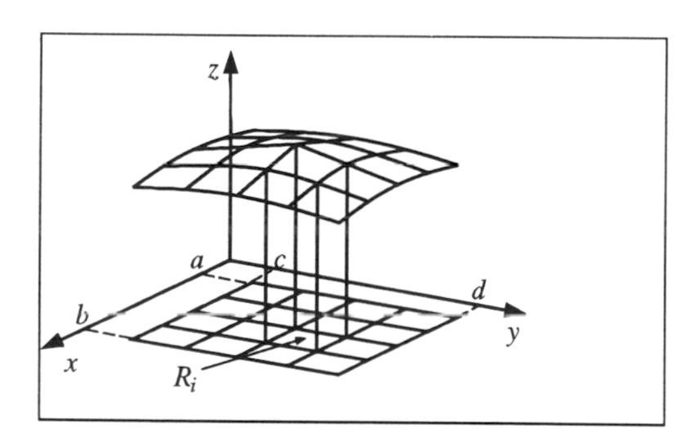

Figure 8.1. Volume under a surface

In a manner similar to the derivation of the definite integral for plane areas we split the rectangular region into smaller rectangles whose sides are parallel to the largest rectangle. The area of each rectangle is denoted by ΔR_i and the volume enclosed between this small rectangle and the surface is *approximately* equal to the cuboid $f(x_i, y_i)\Delta R_i$ where $f(x_i, y_i)$ is the function value (or height) at the centre of the ith rectangle. By summing over all of these small rectangles we have the following estimate of the volume

$$V \approx \sum_{i=1}^{n} f(x_i, y_i) \Delta R_i \ .$$

As the area of these rectangles tends to zero we intuitively expect this estimate of the volume to approach the actual value. If the area of the rectangle is written $R_i = dx\, dy$ then we get the definition

$$\lim_{n \to \infty} \sum_{i=1}^{n} f(x_i, y_i)\, dxdy = \int\int_R f(x, y)\, dxdy \ .$$

Rather crudely we have obtained a definition of the double integral which we have interpreted as a volume. The difficulty facing us now is: how do we evaluate this integral?

The evaluation depends a great deal on the specification of the region in the x, y plane and on the form of the function $f(x, y)$. Without proof we state the result:

Definition 8.1

If the region R is a rectangle and the function $f(x, y)$ can be written as the product of a function of x only and a function of y only, say $f(x, y) = g(x)h(y)$, then the double integral is the product of two single integrals

$$\int\int_R f(x, y) dxdy = \left(\int_a^b g(x) dx \right) \left(\int_c^d h(y) dy \right) \ .$$

Problem 8.1

Evaluate the double integral

$$\int\int x^2 y^2 dxdy$$

over the rectangular region R defined by $1 \le x \le 4$, $2 \le y \le 6$.

Solution 8.1

The limits of integration on x are $x = 1$ and $x = 4$ while the limits on y are $y = 2$ and $y = 6$. Therefore

$$\begin{aligned} \int\int_R x^2 y^2 dxdy &= \int_2^6 \int_1^4 x^2 y^2 dxdy = \int_2^6 \left[\int_1^4 x^2 y^2 dx \right] dy \\ &= \int_2^6 \left[\frac{x^3}{3} y^2 \right]_1^4 dy = \int_2^6 \frac{63}{3} y^2 dy = \frac{63}{3} \left[\frac{y^3}{3} \right]_2^6 = \frac{13104}{9} \ . \end{aligned}$$

This indicates that the volume enclosed between the defined rectangle in the x, y plane and the surface $z = x^2 y^2$ is $\frac{13104}{9}$ units^3.

Non-rectangular regions

Suppose the region R, illustrated in figure 8.2, in the x, y plane is bounded by the two curves $y = g_1(x)$ and $y = g_2(x)$ which intersect at $x = a$ and $x = b$.

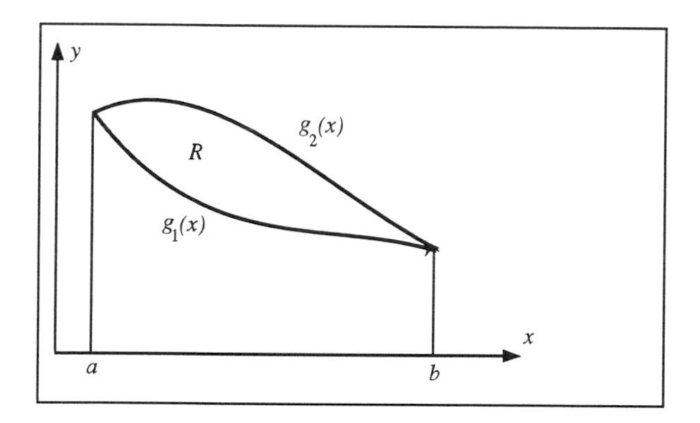

Fig. 8.2. Non-rectangular region R

The double integral over the region R is evaluated as

$$\int\int_R f(x, y)dx\,dy = \int_a^b \left(\int_{g_1(x)}^{g_2(x)} f(x, y)dy\right) dx$$

where the brackets have been included to emphasise that f must be integrated with respect to y first, treating x as a constant, and then this result is integrated with respect to x from $x = a$ to $x = b$. This type of integral is called an *iterated* or *repeated* integral. Note that it is the region in the x,y plane, and not the integrand $f(x, y)$ that determines the limits.

For example, suppose we wish to evaluate the double integral of a function $f(x, y)$ over the region R bounded by the ellipse $x^2 + 4y^2 = 4$ using cartesian coordinates, as shown in figure 8.3 .

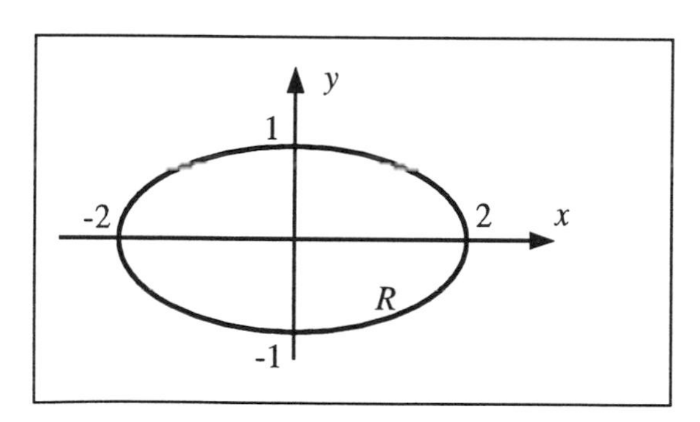

Fig. 8.3. Bounding ellipse

We need the limits on the x and y integrations. If we integrate over x first and y second, then the limits on x come from the boundary of R. We have

$$x^2 = 4(1-y^2), \qquad x = \pm 2\sqrt{1-y^2},$$

so the upper and lower limits on x are $2\sqrt{1-y^2}$ and $-2\sqrt{1-y^2}$ respectively. The limits on y are when $x=0$, i.e. $4y^2 = 4$, so $y = \pm 1$. Thus

$$\int\int_R f(x,y)dx\,dy = \int_{-1}^{1}\left(\int_{-2\sqrt{1-y^2}}^{2\sqrt{1-y^2}} f(x,y)\,dx\right)dy\,.$$

Problem 8.2

Evaluate the double integral

$$\int\int_R x^{\frac{1}{2}}y\,dx\,dy$$

where R is the region bounded by $x>0$, $y>x^2$, $y<2-x^2$.

Solution 8.2

This region is the area in the first quadrant between the quadratics $y=x^2$ and $y=2-x^2$. The quadratics intersect at $x=1$ and therefore the limits on x are $x=0$ and $x=1$.

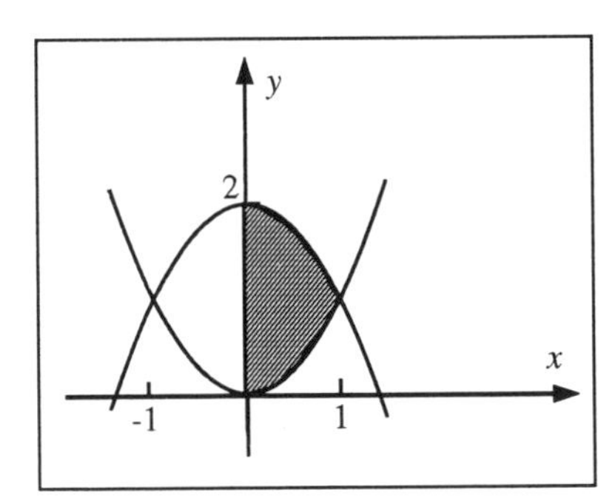

Fig. 8.4. Region of integration

$$\begin{aligned}\int\int_R x^{\frac{1}{2}}ydxdy &= \int_0^1\int_{x^2}^{2-x^2} x^{\frac{1}{2}}y\,dy\,dx = \int_0^1 x^{\frac{1}{2}}\left[\frac{y^2}{2}\right]_{x^2}^{2-x^2} dx \\ &= \frac{1}{2}\int_0^1 x^{\frac{1}{2}}(4-4x^2)dx = 2\int_0^1 x^{\frac{1}{2}} - x^{\frac{5}{2}}dx \\ &= 2\left[\frac{2}{3}x^{\frac{3}{2}} - \frac{2}{7}x^{\frac{7}{2}}\right]_0^1 = \frac{16}{21}\,.\end{aligned}$$

Surface area

Another application of double integrals is in the calculation of surface area. The area A of the surface $z = f(x, y)$ for (x, y) in a region R is *defined* by

$$A = \int\int_R \sqrt{1 + \left(\frac{\partial f}{\partial x}\right)^2 + \left(\frac{\partial f}{\partial y}\right)^2}\, dx\, dy\,,$$

provided the partial derivatives are continuous over R.

Problem 8.3

Find the surface area of $z = x + \sqrt{2}y + 1$ in a region bounded by $y = x^2$ and $y = 1$.

Solution 8.3

The intersection of the y functions are given by $x^2 = 1$ and so the limits

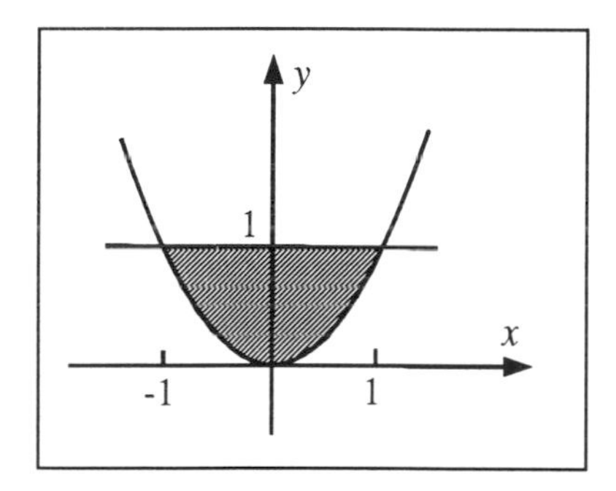

Fig. 8.5. Region of integration

on x are $x = \pm 1$. Next find the partial derivatives of the function z.

$$\frac{\partial z}{\partial x} = 1, \qquad \frac{\partial z}{\partial y} = \sqrt{2}\,.$$

Then the surface area A is given by

$$\begin{aligned} A &= \int_{-1}^{1}\int_{x^2}^{1} \sqrt{1 + \left(\frac{\partial r}{\partial x}\right)^2 + \left(\frac{\partial r}{\partial y}\right)^2}\, dy\, dx \\ &= \int_{-1}^{1}\int_{x^2}^{1} \sqrt{1+1+2}\, dy\, dx = \int_{-1}^{1}\int_{x^2}^{1} 2\, dy\, dx \\ &= \int_{-1}^{1} [2y]_{x^2}^{1}\, dx = \int_{-1}^{1} 2 - 2x^2 dx = \left[2x - \frac{2}{3}x^3\right]_{-1}^{1} = \frac{8}{3}\,. \end{aligned}$$

This example was particularly simple because the function was a plane.

If the integrand is unity what does $\int\int_R dx\,dy$ represent?
The integral represents the volume enclosed between the region R in the x, y plane, $z = 0$, and the plane $z = 1$. As the volume has unit height everywhere the integral is also a measure of the area of the region R.

8.1.1 Change of variable in double integrals

We wish to make a change of variable $x = x(u, v)$, $y = y(u, v)$, which, in certain cases, makes it easier to evaluate double integrals of the form

$$\int\int f(x, y)\ dx\,dy\ .$$

Suppose we can draw curves of constant u and v values in the x, y plane.

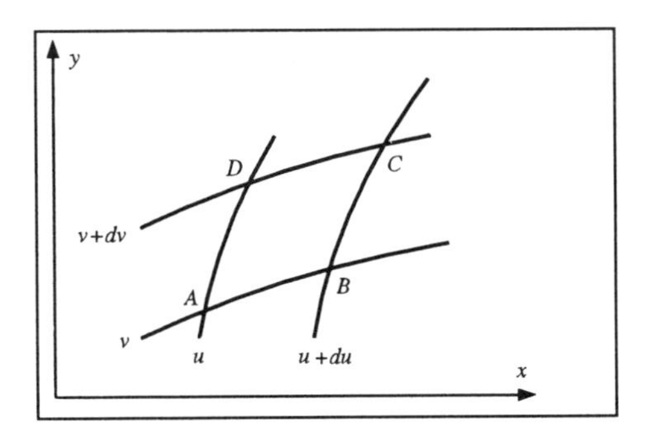

Fig. 8.6. Constant u and v representation

Changes in x and y due to changes in u and v are represented by differentials

$$dx = \frac{\partial x}{\partial u}du + \frac{\partial x}{\partial v}dv \qquad dy = \frac{\partial y}{\partial u}du + \frac{\partial y}{\partial v}dv\ .$$

The points A, B, C and D in figure 8.6 have the coordinates:

	(u, v) coordinates	(x, y) coordinates
A	(u, v)	(x, y)
B	$(u + du, v)$	$(x+\frac{\partial x}{\partial u}du, y+\frac{\partial y}{\partial u}du)$
C	$(u + du, v + dv)$	$(x+\frac{\partial x}{\partial u}du+\frac{\partial x}{\partial v}dv, y+\frac{\partial y}{\partial u}du+\frac{\partial y}{\partial v}dv)$
D	$(u, v + dv)$	$(x+\frac{\partial x}{\partial v}dv, y+\frac{\partial y}{\partial v}dv)$

In general $ABCD$ will not be a rectangle because we cannot say that lines of constant u and v intersect at right angles. However, to a first approximation $ABCD$ is a parallelogram whose area can be found using the vector product $\mathbf{AB} \times \mathbf{AD}$. In this case we have

$$\begin{aligned}
\mathbf{AB} &= \frac{\partial x}{\partial u}du\mathbf{i} + \frac{\partial y}{\partial u}du\mathbf{j} = \left(\frac{\partial x}{\partial u}\mathbf{i} + \frac{\partial y}{\partial u}\mathbf{j}\right) du, \\
\mathbf{AD} &= \frac{\partial x}{\partial v}dv\mathbf{i} + \frac{\partial y}{\partial v}dv\mathbf{j} = \left(\frac{\partial x}{\partial v}\mathbf{i} + \frac{\partial y}{\partial v}\mathbf{j}\right) dv, \\
\mathbf{AB} \times \mathbf{AD} &= \begin{vmatrix} \mathbf{i} & \mathbf{j} & \mathbf{k} \\ \frac{\partial x}{\partial u}du & \frac{\partial y}{\partial u}du & 0 \\ \frac{\partial x}{\partial v}dv & \frac{\partial y}{\partial v}dv & 0 \end{vmatrix} = \begin{vmatrix} \frac{\partial x}{\partial u} & \frac{\partial y}{\partial u} \\ \frac{\partial x}{\partial v} & \frac{\partial y}{\partial v} \end{vmatrix} du\, dv\, \mathbf{k}, \\
\text{Area} &= |\mathbf{AB} \times \mathbf{AD}| = |J|\, du\, dv\ ,
\end{aligned}$$

where J is the second order determinant and is called the *Jacobian* of (x, y) with respect to (u, v). The modulus signs are necessary to ensure a positive element of area. A common notation is

$$J = \frac{\partial(x, y)}{\partial(u, v)}\ .$$

Problem 8.4

The following double integral is to be evaluated using polar coordinates, where the region R is the semicircle of unit radius lying above the x-axis

$$\int\int_R x^2 + y^2\ dx\, dy\ .$$

Solution 8.4

For polar coordinates we have

$$\begin{aligned}
x &= r\cos\theta, & \frac{\partial x}{\partial r} &= \cos\theta, & \frac{\partial x}{\partial \theta} &= -r\sin\theta, \\
y &= r\sin\theta, & \frac{\partial y}{\partial r} &= \sin\theta, & \frac{\partial y}{\partial \theta} &= r\cos\theta, \\
J &= \begin{vmatrix} \frac{\partial x}{\partial r} & \frac{\partial x}{\partial \theta} \\ \frac{\partial y}{\partial r} & \frac{\partial y}{\partial \theta} \end{vmatrix} = r(\cos^2\theta + \sin^2\theta) = r\ .
\end{aligned}$$

Thus the element of area equivalent to $dx\, dy$ is $r\, dr\, d\theta$ and the integral becomes

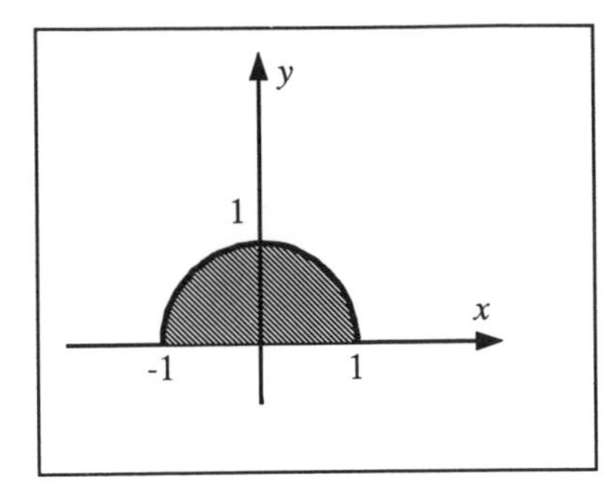

Fig. 8.7. Semicircle of unit radius

$$\begin{aligned}\int\int_R x^2+y^2\,dx\,dy &= \int_0^{\pi}\int_0^1 r^2\,r\,dr\,d\theta \\ &= \int_0^{\pi}\left[\frac{r^4}{4}\right]_0^1 d\theta = \int_0^{\pi}\frac{1}{4}\,d\theta = \frac{\pi}{4}\,.\end{aligned}$$

EXERCISES

Evaluate the following iterated integrals.

8.1. $\int_0^1\int_{x^2}^{x} xy\ dydx$

8.2. $\int_0^1\int_0^{2x^2} xe^y\ dydx$

8.3. Calculate the surface area of the function $z=4x-3y+17$ bounded by the region $x=0$, $y=x$ and $y=1$.

8.2 Triple integrals

Triple integrals are used to find the volumes of three-dimensional shapes, masses and moments of inertia or average values of functions in space. The treatment of triple integrals is very similar to that of double integrals.

The function to be integrated is now $f(x,y,z)$, a function of *three* variables and the region of integration is a *closed* volume V in three-dimensional space. In cartesian coordinates we have

$$\int\int\int_V f(x,y,z)\ dx\ dy\ dz$$

and if $f(x,y,z)=1$ this triple integral measures the volume of V.

For example, if the region is the cuboid for which $1\le x\le 3$, $2\le y\le 6$, $0\le z\le 4$ then the volume of the cuboid is given by

$$\int_0^4 \int_2^6 \int_1^3 dx\, dy\, dz = \int_0^4 \int_2^6 [x]_1^3\, dy\, dz = \int_0^4 \int_2^6 2\, dy\, dz$$
$$= \int_0^4 [2y]_2^6\, dz = \int_0^4 8\, dz = [8z]_0^4 = 32\,.$$

Problem 8.5

Evaluate the triple integral

$$\int_1^3 \int_2^3 \int_1^2 (x - y + z) dx\, dy\, dz\,.$$

Solution 8.5

We integrate with respect to x, y and z in turn substituting the limits.

$$\int_1^3 \int_2^3 \int_1^2 (x - y + z) dx\, dy\, dz = \int_1^3 \int_2^3 \left[\frac{x^2}{2} - yx + zx\right]_1^2 dy\, dz$$
$$= \int_1^3 \int_2^3 \frac{3}{2} - y + z\, dy\, dz = \int_1^3 \left[\frac{3y}{2} - \frac{y^2}{2} + zy\right]_2^3 dz$$
$$= \int_1^3 z - 1\, dz = 2\,.$$

Problem 8.6

A function $f(x, y, z)$ is to be integrated over the region V shown in figure 8.8. The region is bounded by the planes $x = 0$, $y = 0$, $z = 0$ and $x + y + z = 1$. If the integration is to be carried out over z first, then y and finally x determine the limits of integration. Calculate the volume of this region V.

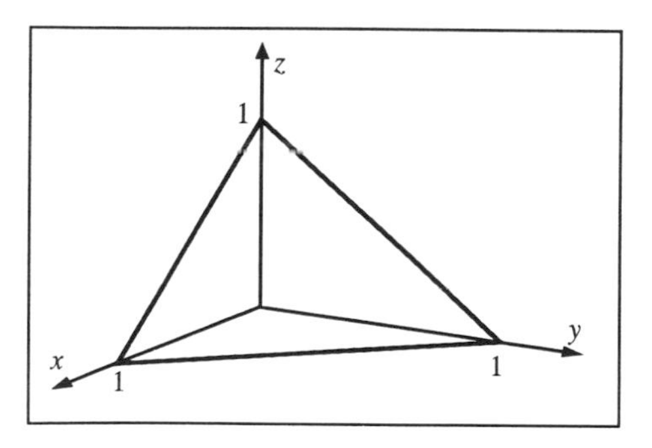

Fig. 8.8. Tetrahedron

Solution 8.6

The lower boundary, or lower limit, on z is the plane $z = 0$. A line drawn from any point on $z = 0$ (which is part of the boundary of V) and parallel to the z-axis will meet the plane $x+y+z = 1$ when $z = 1-x-y$, which is the upper limit on z.

To find the limits of integration on y we keep x constant and consider the x,y plane. The lower limit is $y = 0$ and a line from, and parallel to, this plane will meet the boundary when $x + y = 1$, i.e. $y = 1 - x$ is the upper limit on y. Finally, the limits on x are 0 and 1.
Thus the integral is

$$\int_0^1 \int_0^{1-x} \int_0^{1-x-y} f(x,y,z)\, dz\, dy\, dx \,.$$

The volume of the region corresponds to $f(x,y,z) = 1$.

$$\begin{aligned} \text{Volume} &= \int_0^1 \int_0^{1-x} \int_0^{1-x-y} dz\, dy\, dx \\ &= \int_0^1 \int_0^{1-x} [z]_0^{1-x-y}\, dy\, dx = \int_0^1 \int_0^{1-x} 1-x-y\, dy\, dx \\ &= \int_0^1 \left[y - xy - \frac{y^2}{2}\right]_0^{1-x} dx = \int_0^1 \frac{1}{2}(1-x)^2\, dx \\ &= \left[-\frac{1}{6}(1-x)^3\right]_0^1 = \frac{1}{6}\,. \end{aligned}$$

EXERCISES

8.4. Evaluate the triple integral

$$\int_0^1 \int_{x^2}^{x} \int_0^{xy} dz\, dy\, dx \,.$$

8.5. Find the volume within the circle $x^2 + y^2 = 9$ above the plane $x = 0$ and below the plane $x + z = 4$.

9
Differential Equations

9.1 Introduction

Differential equations appear in many diverse subject areas, most often within mathematical models of processes in physical and engineering systems. This is a direct consequence of the element of change and variability inherent in almost all natural phenomena and physical processes.

The ability to recognise and solve standard categories of differential equations is an extremely useful mathematical skill. Before developing such techniques we shall illustrate some particular applications of differential equations, answer some commonly asked questions and explain some of the notation used in differential equations.

Bungee jumping

A fairly recent phenomenon which appeals to the more adventurous citizens is bungee jumping, which can be modelled using the equation of motion for a mass attached to one end of an elastic string. In this application the modelling is particularly important because accurate estimation of the maximum extension of the cord is crucial to the well-being of the jumper! A simple model of the bungee jumper's motion can be obtained using Newton's second law of motion: *'the product of the mass and the acceleration is equal to the external force'.*

If we neglect air resistance and the energy given up as heat when the bungee cord repeatedly stretches and contracts then the following equation describes the motion of the jumper

$$m\frac{d^2y}{dt^2} = -ky \ .$$

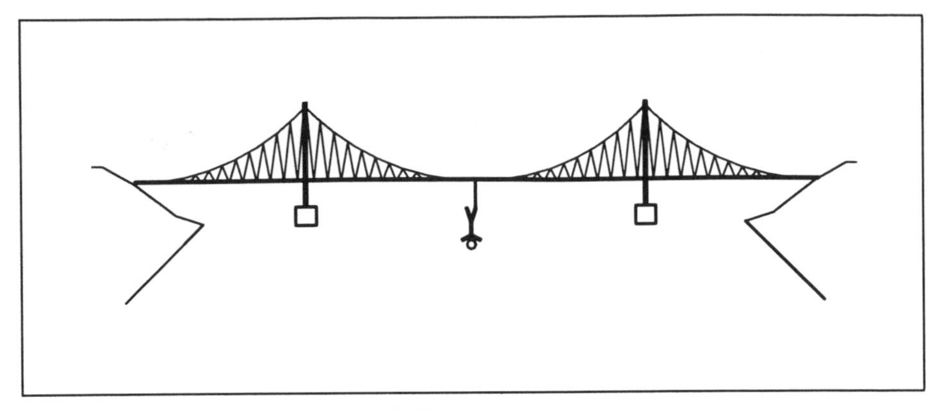

Fig. 9.1. Bungee jumping from a bridge

Here k represents the stiffness of the cord, m is the mass of the jumper and y is the distance from the equilibrium position, the position where the jumper would hang stationary on a stretched cord. The solution is

$$y(t) = y_0 \cos\left(\sqrt{\frac{k}{m}}t\right)$$

where y_0 is the starting position of the jumper relative to the equilibrium position, i.e. the distance from the equilibrium position to the bridge.

Notice that this solution is oscillatory because of the cosine function and therefore could possibly simulate the bouncing motion of the jumper. The biggest failing of this solution is that it predicts that the motion of the jumper has a constant amplitude y_0 and will oscillate for ever! We shall show later that by including resistance terms a more realistic damped motion is predicted.

Blowing the kids' inheritance

A couple reach retirement age with a regular pension of £750 per month and the total value of their house, savings and investments amounts to £180,000. The pension is sufficient to cover all their living expenses and pay for a three-week vacation each year, leaving the £180,000 of assets to accumulate interest for their children to inherit when the couple die.

However, this particular elderly couple want to 'live a little'. They reckon on surviving for another twenty years and want to maximise their spending power by realising their assets and gradually using up the capital. They calculate how much they can withdraw each month by modelling the cashflow as a continuous function rather than as discrete interest payments and withdrawals. The capital is assumed to generate r % interest continuously over each month.

A possible differential equation modelling the amount of capital, $p(t)$, is

$$\frac{dp}{dt} = rp - W \ .$$

On the right hand side the first term, rp, indicates an increase in capital due to accruing interest while the term, $-W$, reflects the reduction in capital due to withdrawals.

Later in the chapter we derive the solution to this equation but for the moment we simply state the solution as

$$p(t) = \left(p_0 - \frac{W}{r}\right) e^{rt} + \frac{W}{r} \ .$$

In this example the initial capital, p_0, is £180,000, the final time is 20 years or 240 months and ideally the couple want $p(240) = 0$. So given the monthly interest rate, r, we calculate the monthly withdrawals to be

$$W = \frac{p_0 r e^{240r}}{e^{240r} - 1} \ .$$

If the interest rate is equivalent to 0.5% per month the couple would be able to withdraw approximately £1288 per month to supplement their pension.

Although this example is a little tongue in cheek the equations are a reasonable approximation to the case where instead of withdrawals a person makes regular contributions to a savings plan to build up a lump sum for retirement.

Chemical extraction

A method of selectively separating or removing particular species from chemical mixtures is liquid–liquid extraction. For example, methanol mixes easily with both water and toluene but given a choice it will mix more readily with water.

A typical extraction process to separate the components of a methanol–toluene mixture is to pass droplets of distilled water through the mixture. Because water is more dense than the methanol–toluene mixture the water droplets fall through the mixture, absorbing the methanol as they go. The droplet speed depends on many factors, the most important being shape, size and the relative concentrations of the chemical species. The speed v (positive downwards) of the droplet can be modelled crudely by

$$\frac{dv}{dt} = \left(1 - \frac{\rho_0}{\rho}\right) g - R_D$$

where ρ is the density of the droplet and ρ_0 the density of the mixture. Notice that while $\rho > \rho_0$, the droplet is more dense than the surrounding mixture and the term in brackets represents a positive acceleration tending to increase the

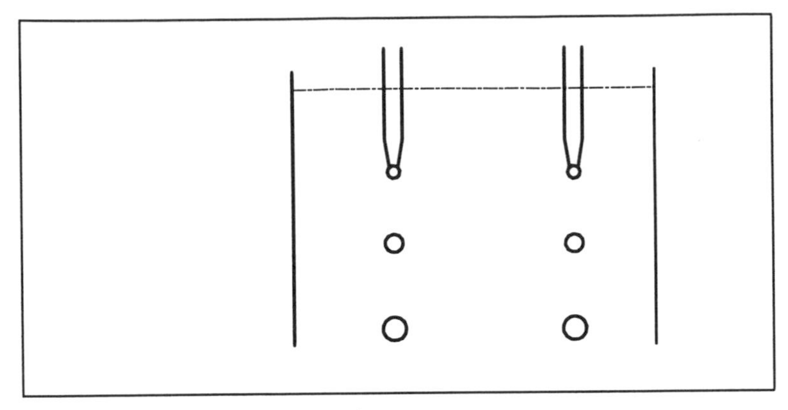

Fig. 9.2. Liquid–liquid extraction

downward speed. As the droplet falls there will be some resistance R_D exerted by the surrounding liquid, referred to as drag, which will vary according to the speed and size of the droplet.

Some common questions answered

What is a differential equation?
From the illustrative examples it should be apparent that a differential equation is simply an equation that contains one or more derivative terms.

Are all differential equations solvable?
Only particular categories of differential equation can be solved analytically using a few specific techniques. A greater range can be solved using numerical methods but these are not covered in this chapter.

How are differential equations categorised?
The types of differential equations are described by their *degree*, whether they are *linear* or *non-linear* and by their *order*.

In this chapter we only consider differential equations involving two variables: one independent variable and one dependent variable.

The following three examples will be used to illustrate the notation.

$$\frac{dy}{dx} = x^2 y \tag{9.1}$$

$$\left(\frac{dP}{dT}\right)^2 = -k(T - T_0) \tag{9.2}$$

$$\frac{d^2 v}{dt^2} + 3\frac{dv}{dt} = 3v^2 - 6\cos t \tag{9.3}$$

When solving the three differential equations given above we are attempting to find the functions $y(x)$, $P(T)$ and $v(t)$. In these examples the *independent* variables are x, T and t respectively and the corresponding *dependent* variables are y, P and v.

The *degree* of a differential equation is the power on the highest derivative. The highest derivative in equations (9.1) to (9.3) are 1, 1 and 2 respectively. Looking *only* at the power on these derivative terms we find the degree of equations (9.1) to (9.3) is 1, 2 and 1 respectively.

Differential equations are further categorised as being *linear* or *non-linear*. Linearity or otherwise only applies to the dependent variable and its derivatives. For an equation to be linear the power on the dependent variable and its derivatives must be one or zero.

The *order* of a differential equation is the highest derivative in the equation. In equations (9.1) and (9.2) the only derivative is a first derivative and so both equations are order one. The highest derivative in equation (9.3) is a second derivative and so its order is two.

For example, the following differential equation is second order and degree one.

$$a\frac{d^2y}{dx^2} + b\frac{dy}{dx} + cy = f(x)$$

It is linear if a, b and c are constants or functions of x only.

Returning to the examples above, equation (9.1) is linear. However, equation (9.2) is non-linear because of the power of two on the derivative. Equation (9.3) is also non-linear because the term $3v^2$ is non-linear in v.

EXERCISES

For the following differential equations write down the order and degree of each, then state whether each is linear or non-linear.

9.1. $\dfrac{dy}{dx} = (2x+5)^3$

9.2. $\dfrac{d^3y}{dx^3} - x\dfrac{dy}{dx} = \cos\left(3y + \frac{\pi}{3}\right)$

9.3. $\dfrac{d^2P}{dr^2} = e^{3P}$

9.4. $\dfrac{dg}{dt} = \dfrac{g}{t}$

The remainder of this chapter illustrates a range of standard methods for solving first and second order differential equations. The first order equations cover a number of categories but the second order equations are all linear.

9.2 First order differential equations

9.2.1 Equations of the form $\frac{dy}{dx} = f(x)$

The simplest category of differential equations to solve are those of the form

$$\frac{dy}{dx} = f(x) \tag{9.4}$$

where the rate of change of y is a function of the independent variable x *only.*

This type of differential equation gives a gentle introduction to the topic but, as we shall see shortly, such simple equations rarely give a realistic mathematical model of physical phenomena.

When we 'solve' the differential equation (9.4) we are trying to find a function (or family of functions) $y(x)$ which, when differentiated, gives $f(x)$. This is accomplished by direct integration

$$y(x) = \int f(x)\ dx. \tag{9.5}$$

There are functions, $f(x)$, which are extremely difficult or even impossible to integrate analytically. Such functions will not be used in this chapter because they do not help in understanding the nature of differential equations.

Problem 9.1

Find the general solution of the differential equation

$$\frac{dy}{dx} = 2x. \tag{9.6}$$

Solution 9.1

This is of the same type as equation (9.4), with $f(x) = 2x$, and so the solution is given by

$$y(x) = \int 2x\ dx = x^2 + C \tag{9.7}$$

where C is a constant of integration.
It is easy to verify that differentiating both sides of equation (9.7) returns the differential equation (9.6).

Note: it is worth remembering that a solution can always be checked by substituting it back into the original differential equation.

Equation (9.7) is called a *general solution* because it contains an arbitrary constant, C, and satisfies the differential equation for *all* values of C.

Figure 9.3 shows some of the family of solutions generated by equation (9.7). In this example the difference between the solutions is the displacement of the parabola in the y-coordinate direction. A *particular solution* is one with

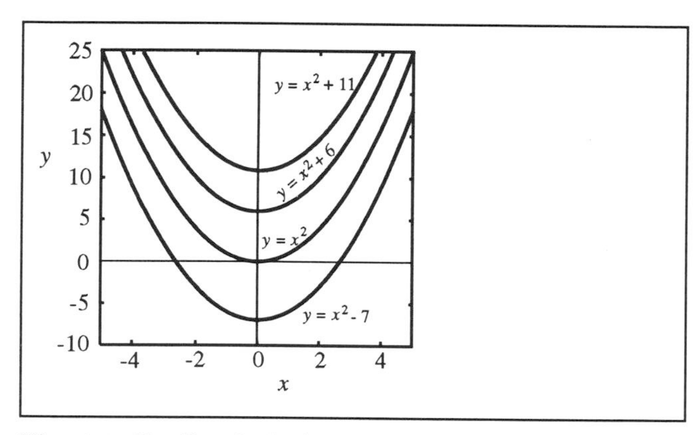

Fig. 9.3. Family of solutions to Problem 9.1

a specific value of the constant C. To find the value of C we require one further piece of information, usually an *initial condition*, relating a particular value of the dependent variable to a particular value of the independent variable, i.e. we select only one of the family of solutions.

The initial condition is usually written in the form

$$y(x_0) = y_0,$$

meaning that $y = y_0$ when $x = x_0$. This terminology was adopted because the independent variable is often time. The initial condition then defines the situation at a fixed instant and the particular solution describes the subsequent behaviour of the dependent variable.

For example, if, for the general solution given by equation (9.7), we have the extra condition that $y = 1$ when $x = 0$, then we can calculate a value of C and hence get a particular solution which satisfies both the differential equation and the initial condition $y(0) = 1$.
In Problem 9.1 the general solution was

$$y(x) = x^2 + C,$$

and substituting the condition $y(0) = 1$ gives

$$1 = 0 + C, \qquad \text{i.e. } C = 1.$$

Therefore the particular solution is

$$y(x) = x^2 + 1\,.$$

We now look at two more straightforward examples before considering some applied problems.

Problem 9.2

Find the particular solution which satisfies the following differential equation and initial condition.

$$\frac{dy}{dx} = \frac{1}{x^2 + 9}, \qquad y(0) = \frac{\pi}{3}. \tag{9.8}$$

Solution 9.2

The right-hand side of equation (9.8) is only a function of x, therefore the general solution is found by direct integration

$$y(x) = \int \frac{1}{x^2 + 9}\, dx.$$

This is a particular example of the standard integral

$$\int \frac{1}{x^2 + a^2}\, dx = \frac{1}{a}\tan^{-1}\left(\frac{x}{a}\right) + C.$$

Comparing the two integrals we see that $a = 3$ and so

$$y(x) = \frac{1}{3}\tan^{-1}\left(\frac{x}{3}\right) + C, \tag{9.9}$$

where C is a constant of integration.
The initial condition, $y(0) = \frac{\pi}{3}$, fixes the value of C. Substituting this condition into the general solution, equation (9.9), gives

$$\frac{\pi}{3} = 0 + C, \qquad \text{i.e.} \qquad C = \frac{\pi}{3}.$$

The particular solution is then

$$y(x) = \frac{1}{3}\tan^{-1}\left(\frac{x}{3}\right) + \frac{\pi}{3}. \tag{9.10}$$

It is quite usual to drop the explicit x dependence and write y instead of $y(x)$ so equation (9.10) could be written as

$$y = \frac{1}{3}\tan^{-1}\left(\frac{x}{3}\right) + \frac{\pi}{3}.$$

Problem 9.3

Find the particular solution which satisfies the following differential equation and initial condition

$$\frac{dP}{dt} = \mathrm{e}^{2t}, \qquad P(0) = 3 \ .$$

Solution 9.3

Again the general solution is found by direct integration

$$P = \int \mathrm{e}^{2t}\, dt = \frac{1}{2}\mathrm{e}^{2t} + C \ . \tag{9.11}$$

The value of C is obtained by substituting the initial condition $P(0) = 3$:

$$3 = \frac{1}{2}\mathrm{e}^{0} + C \ \Rightarrow \ C = \frac{5}{2} \ .$$

The particular solution

$$P = \frac{1}{2}\left(\mathrm{e}^{2t} + 5\right)$$

satisfies both the differential equation and the initial condition.

The three examples solved so far have been exercises in applying a mathematical technique with no need to consider how sensible the solution is. We now apply this technique to some simple applications and in the final problem we illustrate the need to question whether the results obtained seem plausible.

Problem 9.4

A flywheel with moment of inertia I rotates so smoothly about its axis that friction can be neglected. If a constant torque T is applied then the differential equation for the angular velocity ω is

$$I\frac{d\omega}{dt} = T, \qquad \omega(0) = \omega_0,$$

where $\omega_0 > 0$ is the initial angular velocity of the flywheel. Solve this equation to find $\omega(t)$ and interpret the result.

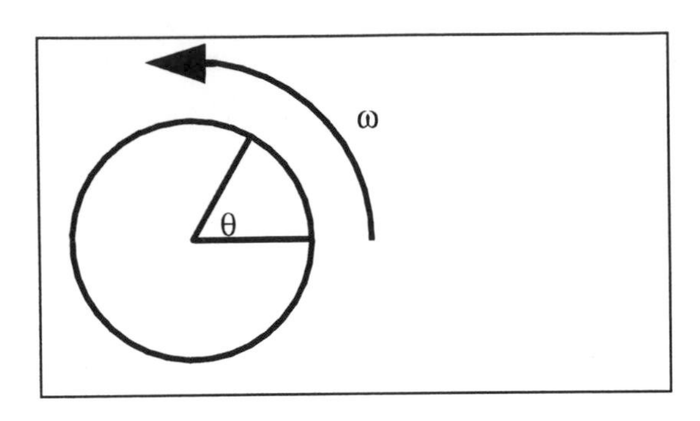

Fig. 9.4. A flywheel rotating about its axis

Solution 9.4

First of all rearrange the equation to get

$$\frac{d\omega}{dt} = \frac{T}{I} .$$

and then integrate to get ω:

$$\omega = \int \frac{T}{I}\, dt = \frac{T}{I} t + C .$$

The value of C is found using the initial condition

$$\omega_0 = 0 + C, \quad \text{i.e.} \quad C = \omega_0 ,$$

and so the particular solution is

$$\omega = \omega_0 + \frac{Tt}{I}$$

The interpretation is straightforward for a constant T. If $T = 0$ then the flywheel continues to rotate at its initial velocity, ω_0. If $T > 0$ the flywheel accelerates and if $T < 0$ the flywheel slows down. In particular, if $T < 0$ then its angular velocity will be zero at time $t = -\frac{I\omega_0}{T}$ after which time the flywheel's velocity becomes negative, i.e. it begins to rotate in the reverse direction.

Problem 9.5

A car is travelling at a constant speed, v, of 25 m/s when the brakes are suddenly applied. The car then slows down according to the equation

$$v = 25 - \frac{10}{3} t$$

where the time t is measured from the instant the brakes are applied. How far does the car travel before it stops?

Solution 9.5

Speed is related to the distance travelled, s, by $v = \frac{ds}{dt}$ and so we replace v by $\frac{ds}{dt}$ in the above equation and solve the differential equation for s.

$$\begin{aligned} \frac{ds}{dt} &= 25 - \frac{10}{3}t \\ s &= \int 25 - \frac{10}{3}t \; dt = 25t - \frac{5}{3}t^2 + C \end{aligned}$$

The initial condition is $s(0) = 0$ because time and hence distance is measured from when the brakes are first applied. Therefore $C = 0$ and the solution for s is

$$s = 25t - \frac{5}{3}t^2$$

To find how far the car travels we need to know what time elapses between applying the brake and the car coming to rest. The car is stopped when $v = 0$:

$$0 = 25 - \frac{10}{3}t \;\Rightarrow\; t = 25 \times \frac{3}{10} = 7.5 \text{ seconds} .$$

Substituting this value of t into the equation for s we get

$$s = 25 \times 7.5 - \frac{5}{3}(7.5)^2 = 93.75 \text{ metres}.$$

Motion in one direction

The equations governing motion of a particle in one direction can be summarised as follows

$$a = \frac{dv}{dt}, \qquad v = \frac{ds}{dt},$$

where at time t the particle's displacement from a fixed point is s, its velocity is v and its acceleration is a.
These are two simple differential equations which can be integrated directly when the acceleration is either constant or a function of time

$$s = \int v \; dt , \qquad v = \int a \; dt .$$

For a constant acceleration a we get

$$\begin{aligned} v &= at + C, \\ \text{then} \qquad s &= \int at + C \; dt = \frac{1}{2}at^2 + Ct + D . \end{aligned}$$

Because we have carried out two integrations we have generated two constants, C and D, which are determined by initial conditions. The usual notation for mechanics problems is to specify the initial velocity to be u, $v(0) = u$, and although the initial displacement is often zero we will keep the more general case $s(0) = s_0$. Substituting these conditions yields the values $C = u$, $D = s_0$ and hence the equations of motion for a constant acceleration are

$$\begin{aligned} v &= u + at, \\ s &= s_0 + ut + \frac{1}{2}at^2 . \end{aligned}$$

We now use these ideas to model a falling raindrop.

Problem 9.6

A thunder cloud is passing 2 km above a small town. In the cloud, raindrops 5mm in diameter are formed and released with zero vertical velocity. If we neglect the effects of air resistance, evaporation to the surrounding air and assume the only important force acting is gravity, at what speed do the raindrops strike the ground?

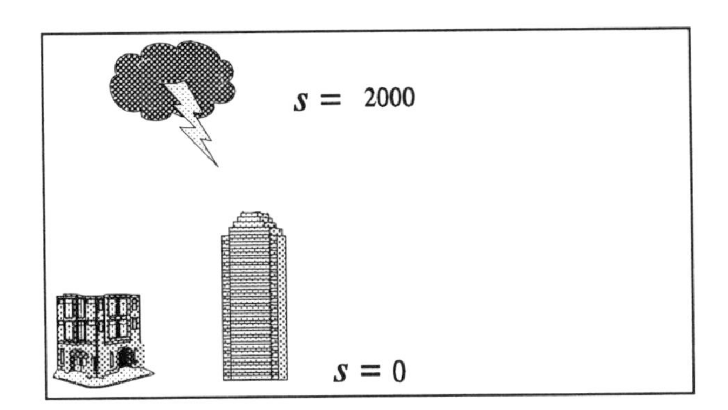

Fig. 9.5. Thundercloud 2 km above town

Solution 9.6

We take ground level as our reference $s = 0$ and define positive motion to be upwards. Instead of a our acceleration is $-g$, the acceleration due to gravity. We have

$$v = u - gt, \qquad s = s_0 + ut - \frac{1}{2}gt^2 .$$

The initial conditions are $u = 0$ and $s_0 = 2000$ and so we have

$$\begin{aligned} v &= -gt, \\ s &= 2000 - \frac{1}{2}gt^2 . \end{aligned}$$

From the second equation we calculate the time taken for the raindrop to reach the ground, $s = 0$.

$$\frac{1}{2}gt^2 = 2000 \Rightarrow t = \sqrt{\frac{4000}{g}}.$$

Taking $g = 9.81 \text{ m/s}^2$ gives $t \approx 20.19$ seconds.

Finally, the speed of the raindrop when it hits the ground is

$$v = -9.81 \times 20.19 = -198.09 \text{ m/s}.$$

From observation and measurement it is known that the terminal velocity of a 5 mm diameter raindrop is little more than 9 m/s! How can we explain the large discrepancy between measurement and prediction?

The above solution is correctly calculated, so let us look at some of the shortcomings of the problem description. The raindrop has not been modelled properly – the above solution does not take into account the drop size. The most important factor we have neglected is the drag force representing air resistance. Including this yields a differential equation which cannot be solved using direct integration but can be solved using techniques described in later sections.

Summary

In this section we have found the general solution to differential equations using direct integration. This technique can *only* be used if the right-hand side is a function of the independent variable only

If an initial condition is also specified then a particular solution can be found which satisfies both the differential equation and the initial condition.

For a differential equation of the form

$$\frac{dy}{dx} = f(x)$$

the general solution is

$$y(x) = \int f(x) \; dx.$$

The general solution contains one arbitrary constant.

Differential equations of this type often fail to model physical phenomena because we are forced to neglect important features in the mathematical model describing the process. For example, by neglecting air resistance to the falling raindrop, we generated predictions contradicting observation.

EXERCISES

For exercises 9.5 to 9.11 find the particular solution satisfying the differential equation and initial condition.

9.5. $\frac{dy}{dx} = x^{\frac{3}{2}}$, $y(0) = 1$

9.6. $\frac{dy}{dx} - x^3 = 1$, $y(2) = 6$

9.7. $\frac{dy}{dx} - e^x = 0$, $y(0) = 1$

9.8. $\frac{dy}{dx} = \cos x$, $y\left(\frac{\pi}{2}\right) = 2$

9.9. $\frac{di}{dt} = 3\sin\left(4t + \frac{\pi}{3}\right)$, $i(0) = 0$

9.10. $\frac{dy}{dx} = \cos\left(3x + \frac{\pi}{3}\right)$, $y(0) = 0$

9.11. $\frac{dy}{dx} = (2x + 5)^3$, $y(0) = 0$

9.12. A metal sphere expands when heated such that the radius, r, increases at a rate of 0.005 cm/s. If the radius of the sphere is originally 2.1 cm express the radius as a function of time, t. How long will it take before the volume, $V = \frac{4}{3}\pi r^3$, of the sphere has increased by 10%?

9.13. A disc is rotating with time-varying angular velocity, $\frac{d\theta}{dt}$, given by

$$\frac{d\theta}{dt} = 1.5\cos\left(\frac{t}{6}\right).$$

Solve this differential equation for the angular displacement, θ, given the initial condition $\theta = 0$ radians at time $t = 0$ seconds.

9.14. The atmospheric pressure P pascals at a height h metres above sea-level is given by the solution of the differential equation

$$\frac{dP}{dh} = -P_0 k e^{-kh},$$

where the constant $k = 1.25 \times 10^{-4}$ and P_0 is the pressure at sea-level. Solve this equation for P. If $P_0 = 1.01 \times 10^5$ pascals what is the pressure 3000 m above sea-level?

9.15. The rate of change of an electrical voltage E volts with time t seconds is given by

$$\frac{dE}{dt} = 1700\pi\cos(100\pi t) - 3600\pi\sin(100\pi t) .$$

Solve this equation to find $E(t)$ if $E(0) = 45$ volts.

9.16. A car is travelling at a constant speed, v, of 40 m/s when the brakes are suddenly applied. The car then slows down according to

$$v = 40 - 5t,$$

where the time t is measured from the instant the brakes are applied. Replace v by $\frac{ds}{dt}$ in the above equation and solve the resulting differential equation for the distance travelled s. How far does the car travel before it stops?

9.2.2 Equations of the form $\frac{dy}{dx} = ky$

We now consider differential equations of the form of equation (9.12) where k is a constant, $k \neq 0$, and so the right hand side is linear in y.

$$\frac{dy}{dx} = ky \tag{9.12}$$

The general solution to any first order differential equation involves only one arbitrary constant and so if we can find a function with one arbitrary constant which satisfies the differential equation then we must have the general solution.

Equation (9.12) states in words that '*we are seeking a function y whose derivative is a multiple of itself*'. Considering the derivatives of the standard functions, this is the unique characteristic of the exponential function.

Suppose then we try the function $y = Ae^{mx}$, where A is an arbitrary constant. First we write equation (9.12) in the slightly different form

$$\frac{dy}{dx} - ky = 0. \tag{9.13}$$

Substituting $y = Ae^{mx}$ and its derivative into equation (9.13) gives

$$Ame^{mx} - kAe^{mx} = Ae^{mx}(m - k) = 0\ .$$

For an arbitrary constant A this equation is satisfied only if $m = k$. Hence

$$y = Ae^{kx}$$

represents the general solution to the linear differential equation (9.12).

We can also derive this solution from first principles. We collect all terms involving y on the left hand side and all terms involving x on the right hand side

$$\frac{dy}{y} = k\ dx$$

but we must exclude $y = 0$ to avoid a division by zero.
If we now integrate both sides we obtain

$$\int \frac{dy}{y} = \int k \, dx \, .$$

The integration is straightforward

$$\ln |y| = kx + C.$$

This is not the ideal form of the general solution. We would prefer $y = \ldots$ rather than $\ln y = \ldots$. This can be achieved by taking exponentials of both sides to give

$$|y| = \mathrm{e}^{kx+C} = \mathrm{e}^{C}\mathrm{e}^{kx} \;\Rightarrow\; y = \pm\mathrm{e}^{C}\mathrm{e}^{kx}$$

noting that $\mathrm{e}^{C} \neq 0$ for all finite values of C. Writing $A = \pm\mathrm{e}^{C}$ gives the neater form

$$y(x) = A\mathrm{e}^{kx} \tag{9.14}$$

However, as $y = 0$ is a possible solution to the differential equation, it can be incorporated by admitting $A = 0$. Thus the general solution is given by (9.14) where A is an arbitrary constant.

EXERCISES

Using equation (9.13) as a guide, identify the value of k and write down the general solution of the following differential equations.

9.17. $\dfrac{dy}{dx} = 5y$

9.18. $\dfrac{dy}{dx} = -2y$

9.19. $\dfrac{dy}{dx} + \frac{1}{2}y = 0$

9.20. $\dfrac{dx}{dt} + \frac{1}{4}x = 0$

Growth and decay

Differential equations such as equation (9.12) often occur in problems of growth and decay where the independent variable is time, t, and so

$$y(t) = A\mathrm{e}^{kt}. \tag{9.15}$$

This solution represents growth or decay depending on the sign of k.

This is illustrated in figure 9.6. If k is positive then as t increases the magnitude of y increases and y is said to grow exponentially. Conversely, if k is negative then as t increases the magnitude of y decreases and y is said to decay exponentially to zero.

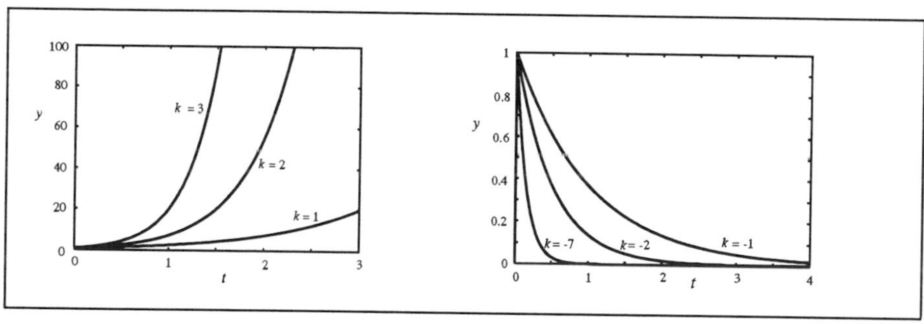

Fig. 9.6. Growth and decay

An initial condition, $y(t_0) = y_0$ where t_0 is the starting point, determines the constant A. If this initial time is zero, i.e. $t_0 = 0$, then the value of the constant A is equal to y_0.

Problem 9.7

The variable y is a function of time (in seconds) and satisfies the equation

$$\frac{dy}{dt} = -\frac{2}{5}y \ . \tag{9.16}$$

If the initial condition is $y(0) = 9$ find the time when $y = 3$, giving this time accurate to 3 decimal places.

Solution 9.7

The solution to this differential equation is of the form given by equation (9.15) with $k = -\frac{2}{5}$ and so we can write down the general solution

$$y = Ae^{-\frac{2}{5}t} \ .$$

The initial condition $y(0) = 9$ gives $9 = Ae^0$, i.e. $A = 9$ and so the particular solution is

$$y = 9e^{-\frac{2}{5}t}.$$

To find the time when $y = 3$ we take logarithms of this particular solution

$$\begin{aligned} 9e^{-\frac{2}{5}t} &= 3, \\ -\frac{2}{5}t &= \ln\left(\frac{1}{3}\right), \\ t &= -\frac{5}{2}\ln\left(\frac{1}{3}\right) = 2.747 \quad \text{(3d.p.)}, \end{aligned}$$

i.e. $y = 3$ after approximately 2.747 seconds.

'Story type' of problems

Applied problems often *appear* more difficult because they are written in the jargon of the subject area and the differential equation is hidden in a 'story'. As far as the solution of the *mathematical problem* is concerned the story can usually be ignored but the context is important for interpreting the solution.

Applied problems often consist of linear differential equations, usually first or second order, whose solutions are exponential functions representing the growth or decay of a quantity. The next two examples illustrate typical problems modelled by first-order differential equations.

Problem 9.8

The rate of growth of bacteria in a culture is directly proportional to the amount of bacteria present. This means that if N is the number of bacteria at time t hours then

$$\frac{dN}{dt} = kN.$$

If the initial number of bacteria present is 2×10^4 find N as a function of t and k. If the number of bacteria doubles in 3 hours find the value of k to 3 decimal places. How many bacteria are present after 4 hours?

Solution 9.8

The first requirement is to express N as a function of t and k. This simply means find the particular solution of the differential equation which satisfies the initial condition $N(0) = 2 \times 10^4$.
The general solution is

$$N(t) = A\mathrm{e}^{kt}$$

and the initial condition gives $A = 2 \times 10^4$ so the solution is

$$N(t) = 2 \times 10^4 \, \mathrm{e}^{kt}.$$

To find k we express the information about doubling the bacteria as

$$\begin{aligned} N(t+3) = 2N(t) \;\Rightarrow\; A\mathrm{e}^{kt+3k} &= 2A\mathrm{e}^{kt}, \\ \mathrm{e}^{3k} &= 2 \end{aligned}$$

and hence $k = 0.231$ (to 3d.p.). Finally, the number of bacteria present after 4 hours is found by substituting $t = 4$ into the particular solution

$$N(4) = 2 \times 10^4 \mathrm{e}^{0.231 \times 4} = 5.039 \times 10^4 \, .$$

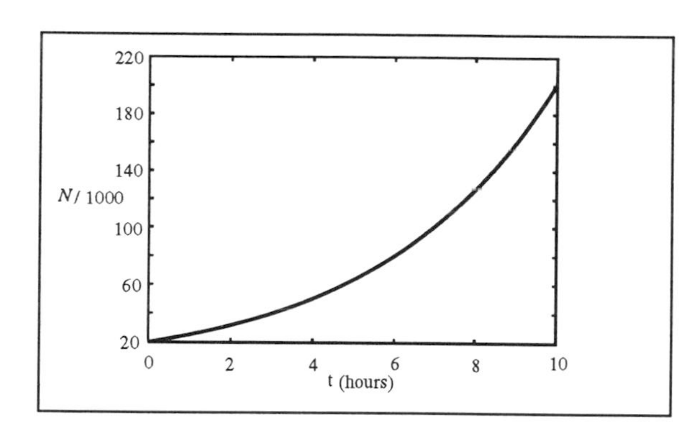

Fig. 9.7. Increase in bacteria

Problem 9.9

In a dilute sugar solution, the rate of decrease of concentration is proportional to the concentration. This means that if c g/cm^3 is the concentration at time t hours, then

$$\frac{dc}{dt} = -kc,$$

where k is a constant. If $c = 0.01$ g/cm^3 at time $t = 0$ and $c = 0.005$ g/cm^3 at time $t = 4$ hours, find k to 3 decimal places and express c as a function of t. What will be the concentration after 10 hours?

Solution 9.9

The general solution is

$$c(t) = Ae^{-kt},$$

and the initial condition $c(0) = 0.01$ gives $A = 0.01$, so

$$c(t) = 0.01\mathrm{e}^{-kt}.$$

To find the value of k we substitute $c(4) = 0.005$.

$$\begin{aligned} 0.01\mathrm{e}^{-4k} &= 0.005, \\ -4k &= \ln 0.5, \end{aligned}$$

so $k = 0.173$ and the solution is

$$c(t) = 0.01\mathrm{e}^{-0.173\,t}.$$

Finally, the concentration after 10 hours is

$$c(10) = 0.01e^{-0.173\times 10} = 1.77 \times 10^{-3}\ \mathrm{g/cm^3}\ .$$

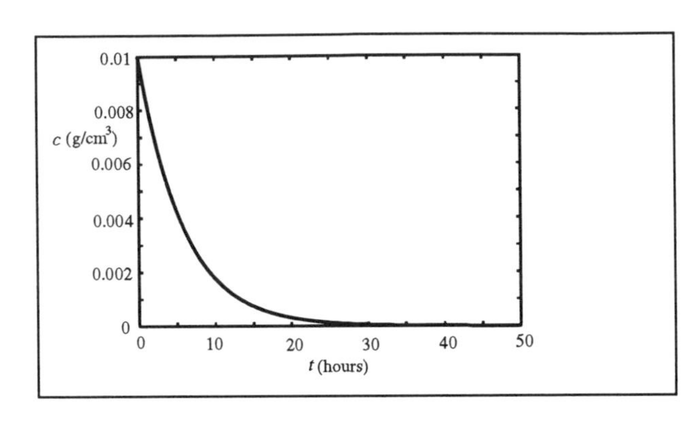

Fig. 9.8. Decrease in concentration

Summary

In this section we have solved differential equations whose right hand sides were a constant multiple of the dependent variable.

If

$$\frac{dy}{dx} = ky$$

the general solution is

$$y = Ae^{kx}.$$

The general solution contains one arbitrary constant, A.

EXERCISES

9.21. Find the particular solutions to the following:

(a) $\dfrac{dy}{dt} = -2y, \quad y(0) = 4$ (b) $\dfrac{dx}{dt} = -3x, \quad x(0) = 2$

(c) $\dfrac{dy}{dt} = -\frac{1}{6}y, \quad y(0) = 7$ (d) $\dfrac{di}{dt} = -\frac{1}{4}i, \quad i(2) = 1$

9.22. A quantity of sugar is dissolved in water. The fraction, F, that remains undissolved after time t, measured in minutes, is assumed to satisfy the differential equation

$$\frac{dF}{dt} = -kF, \qquad k > 0.$$

If it takes one minute to dissolve one quarter of the sugar and k is a constant how long will it take before half the sugar is dissolved?

9.23. The rate of decay of a radioactive material is governed by

$$\frac{dN}{dt} = -\alpha N,$$

where α is a constant and αN is the number of radioactive atoms disintegrating per second. Determine the half-life of radium (the time for N to become one half of its original value) given the decay constant $\alpha = 1.36 \times 10^{-11}$ atoms per second. By assuming a 365-day year express the half-life in years (to the nearest year).

9.24. According to the 1971 census the population, P, of a city was approximately 125,000 and by the time of the 1981 census it had risen to 139,000. If the population is adequately modelled by

$$\frac{dP}{dt} = kP,$$

and k remains constant what will the population be in the year 2001? Give the answer to the nearest thousand.

9.25. A man deposits an amount £P_0 into an account that pays an annual rate of interest r%. If the interest is compounded continuously

$$\frac{dP}{dt} = rP\,.$$

Express the current value of the deposit, P, as a function of time. How long will it take for the initial deposit to treble in value if the annual rate of interest remains constant at 7.5% ?

9.26. The atmospheric pressure P pascals at a height h metres above sea-level is given by the solution of the differential equation

$$\frac{dP}{dh} = -kP,$$

where $k = 1.25 \times 10^{-4}$. If the pressure at sea-level is 1.01×10^{5} pascals calculate the pressure at 5000 m and comment on the value.

9.2.3 Variables separable

The preceding sections described specific types of first order differential equations referred to as *variables separable.* These are differential equations in which all terms involving the independent variable can be collected on one side and all terms involving the dependent variable can be collected on the other.

The equation must be of the form

$$\frac{dy}{dx} = f(x)g(y)\,. \tag{9.17}$$

The first step is to separate the variables, including the differentials

$$\frac{1}{g(y)}\ dy = f(x)\ \ dx\ \ .$$

Now integrate both sides of the equation

$$\int \frac{1}{g(y)}\ dy = \int f(x)\ dx\ .$$

The integration is often straightforward but, occasionally, significant transposition is necessary to express y explicitly in terms of x.

Problem 9.10

Find the general solution to the following differential equation using the technique of separation of variables:

$$\frac{dy}{dx} = y^2 \mathrm{e}^x\ .$$

Solution 9.10

This is the same form as equation (9.17) with $g(y) = y^2$ and $f(x) = \mathrm{e}^x$. First collect the x and y terms on either side of the equation,

$$\frac{1}{y^2}\ dy = \mathrm{e}^x\ dx.$$

Now integrate both sides

$$\int \frac{1}{y^2} dy = \int \mathrm{e}^x dx,$$

to obtain the solution

$$-\frac{1}{y} = \mathrm{e}^x + C,$$

where C is an arbitrary constant.
Finally, rearrange this equation to obtain

$$y = \frac{-1}{\mathrm{e}^x + C}\ .$$

Problem 9.11

Find the general solution to the following differential equation

$$2y(x^2+1)\frac{dy}{dx} = 1\ .$$

Solution 9.11

Separating the x and y terms

$$2y\,dy = \frac{1}{x^2+1}\,dx.$$

Integrating both sides

$$\int 2y\,dy = \int \frac{1}{x^2+1}\,dx$$

leads to the solution

$$y^2 = \tan^{-1} x + C$$

where C is an arbitrary constant. By taking square roots the solution can be expressed explicitly as

$$y = \pm\sqrt{\tan^{-1} x + C}.$$

Note that there are two solutions in this example.

Problem 9.12

Find the particular solution which satisfies the following differential equation and initial condition

$$\frac{1}{2}\frac{dy}{dx} = x + xy \qquad y(0) = 1.$$

Solution 9.12

By factorising the right hand side the differential equation is obviously variables separable

$$\frac{dy}{dx} = 2x(1+y).$$

Now separate the variables,

$$\frac{1}{1+y}\,dy = 2x\,dx.$$

Integrate both sides,

$$\int \frac{1}{1+y}\,dy = \int 2x\,dx,$$

to obtain

$$\ln|1+y| = x^2 + C\ .$$

The constant of integration, C, is found by substituting the initial condition, $y(0) = 1$, thus $C = \ln 2$ and hence

$$\ln(1 + y) = x^2 + \ln 2,$$

which can be rearranged to give the particular solution

$$y = 2e^{x^2} - 1 .$$

As in previous sections solving a differential equation is straightforward when posed in textbook form. When a problem is given in the jargon of the application it suddenly *appears* more difficult. Let us return to the falling raindrop example, problem 9.6 .

Problem 9.13

A thunder cloud is passing 2 km above a town. Raindrops 5 mm in diameter are formed and released with zero vertical velocity. The equation governing the motion of the raindrops is given by

$$\frac{dv}{dt} = g - Ad^2v,$$

where d is the drop diameter (metres), $g = 10$ m/s^2 and $A = 4 \times 10^4$. Express the speed v as a function of time and find its limiting value.

Solution 9.13

As before we specify ground level as $s = 0$ and define positive motion to be upwards. The differential equation is variables separable

$$\frac{dv}{g - Ad^2v} = dt .$$

Integrating both sides

$$\int \frac{dv}{g - Ad^2v} = \int dt,$$

leads to the solution

$$-\frac{1}{Ad^2} \ln \left| g - Ad^2v \right| = t + C .$$

Substitute the initial condition $v(0) = 0$ to find C:

$$-\frac{1}{Ad^2} \ln |g| = C.$$

Therefore the particular solution is

$$
\begin{aligned}
-\frac{1}{Ad^2}\ln\left|1-\frac{Ad^2}{g}v\right| &= t, \\
v &= \frac{g}{Ad^2}\left(1-\mathrm{e}^{-Ad^2t}\right) .
\end{aligned}
$$

The exponential term decays as t increases so the limiting value is

$$v=\frac{g}{Ad^2}=10\ \mathrm{m/s}\ .$$

Summary

In this section we have solved examples of differential equations referred to as variables separable.

If

$$\frac{dy}{dx}=f(x)g(y),$$

then the general solution is found from the solution of

$$\int \frac{1}{g(y)}\,dy=\int f(x)\,dx.$$

The general solution contains one arbitrary constant, A.

EXERCISES

9.27. Find the general solution of the differential equation

$$x^3+(y+1)^2\frac{dy}{dx}=0.$$

9.28. Find the particular solution which satisfies:

$$(1+x^3)\frac{dy}{dx}=x^2y, \qquad y(1)-2.$$

9.29. Solve the first order non-linear differential equation

$$\frac{dy}{dx}=2x\sqrt{1-y^2}, \qquad y(0)=0.5,$$

using the method of separation of variables.

9.30. The voltage, $v(t)$, across the resistor, R, in a series inductance–resistance circuit to which a constant voltage, E volts, is applied, obeys the differential equation

$$L\frac{dv}{dt}=R(E-v)\ .$$

If $v(t) = 0$ when $t = 0$ show that, for $t \geq 0$,

$$v(t) = E(1 - e^{-t/\tau}),$$

where t is time, measured in seconds, and $\tau = \frac{L}{R}$ seconds.
Given that $E = 91$ volts and the value of τ is 0.5 seconds find
(a) the voltage across the resistor after 0.15 seconds,
(b) the time at which $v(t) = 54$ volts.
giving your answers to 3 decimal places.

9.31. A stone falls vertically under gravity against air resistance which is proportional to its velocity. The velocity v is a function of time t and satisfies the relationship

$$\frac{dv}{dt} = g - kv,$$

where g and k are positive constants. If the stone is initially at rest find an explicit expression for the velocity in terms of t, g and k. Show that the limiting value of the velocity is $\frac{g}{k}$.

9.32. Newton's Law of Cooling states that the rate at which a substance cools in moving air is proportional to the difference between the temperature of the substance T and the temperature of the air T_a.

$$\frac{dT}{dt} = K(T - T_a),$$

where K is a positive constant.
If the temperature of the air is 30 ^{o}C and the substance is initially 100 ^{o}C express T as a function of K and t. If it takes 15 minutes for the substance to cool to 70 ^{o}C, find to the nearest tenth of a minute the time it takes for the temperature to drop to 40 ^{o}C.

9.33. The current $i(t)$ in an electrical circuit at time t satisfies

$$Ri + L\frac{di}{dt} - E = 0,$$

where R ohms is the resistance, L henrys is the inductance and E volts is the electromotive force in the circuit. For a particular circuit $R = 300$, $L = 0.8$ and $E = 100$. If $i(0) = 0$ find i as a function of t. What is the value of i after a long period of time?

9.34. A spherical liquid droplet is released from rest into the top of a high vertical tube filled with water. The droplet density is greater than that of water and so the droplet falls down the tube. If the only forces acting on the droplet are gravity, g, and a drag resistance proportional to the droplet speed the differential equation governing its motion can be approximated by

$$\frac{dV}{dt} = g\left(1 - \frac{\rho_0}{\rho}\right) - C_d V,$$

where V is the speed of the droplet, ρ and ρ_0 are the densities of the droplet and water respectively and C_d is the effective drag constant. By posing the problem as

$$\frac{dV}{dt} = A - C_d V, \qquad V(0) = 0,$$

express the speed of the droplet as a function of time, t.
Using the following values, estimate what the limiting speed of the droplet would be if the tube was infinitely high (answer to 4 d.p.).

$$g = 10 \text{ m/s}^2, \quad \rho = 1200 \text{ kg/m}^3, \quad \rho_0 = 1000 \text{ kg/m}^3, \quad C_d = 4.2 \text{ s}^{-1}$$

9.2.4 Integrating factors

Another of the general classifications of differential equation is *linear* differential equations of the form

$$\frac{dy}{dx} + l(x)y = k(x) \,. \tag{9.18}$$

The objective is to find a function called an *integrating factor* such that the left hand side of equation (9.18) can be written in the form

$$\frac{d}{dx}\left(p(x)y\right) = p(x)k(x), \tag{9.19}$$

where $p(x)$ is the integrating factor.

The advantage of this form is that on integrating both sides we obtain

$$p(x)y = \int p(x)k(x)\ dx, \tag{9.20}$$

and so the solution of the original differential equation hinges only on the ability to evaluate the integral on the right hand side of equation (9.20).

To find the form of $p(x)$ we use the product rule to expand the left hand side of equation (9.19) and then compare the result with the modified form of equation (9.18), i.e. we are comparing the following:

$$p(x)\frac{dy}{dx} + p'(x)y = p(x)k(x), \tag{9.21}$$

$$p(x)\frac{dy}{dx} + p(x)l(x)y = p(x)k(x) \,. \tag{9.22}$$

For these to be identical we must have

$$p'(x) = p(x)l(x),$$

and rearranging gives

$$\frac{p'(x)}{p(x)} = l(x).$$

Now we integrate both sides

$$\int \frac{p'(x)}{p(x)}\, dx = \int l(x)\, dx,$$

to obtain

$$\ln p(x) = \int l(x)\, dx\ .$$

The integrating factor is therefore

$$p(x) = \mathrm{e}^{\int l(x)dx}\ . \tag{9.23}$$

It may be possible to solve a differential equation using more than one technique. To illustrate this consider again problem 9.12 but this time we will use the integrating factor method to solve the differential equation.

Problem 9.14

Find the particular solution which satisfies the following differential equation and initial condition.

$$\frac{dy}{dx} - 2xy = 2x, \qquad y(0) = 1.$$

Solution 9.14

Comparing terms with the standard equation (9.18) we see that $k(x) = 2x$ and $l(x) = -2x$. The integrating factor $p(x)$ is given by

$$p(x) = \mathrm{e}^{\int l(x)dx} = \mathrm{e}^{-\int 2x\, dx} = \mathrm{e}^{-x^2}.$$

Note that we neglect the constant of integration. Retaining it would simply multiply every term in the differential equation by a constant which could then be cancelled.

This problem can now be written in the form:

$$\begin{aligned} \frac{d}{dx}\left(\mathrm{e}^{-x^2} y\right) &= 2x\mathrm{e}^{-x^2}, \\ \mathrm{e}^{-x^2} y &= \int 2x\mathrm{e}^{-x^2}\, dx = -\mathrm{e}^{-x^2} + C\ . \end{aligned}$$

Hence the general solution is

$$y = Ce^{x^2} - 1\,.$$

Substituting $y(0) = 1$ gives $C = 2$ and so the particular solution is

$$y = 2e^{x^2} - 1,$$

as found in solution 9.12.

Problem 9.15

Find the particular solution which satisfies the following differential equation and initial condition for $x > 0$

$$\frac{dy}{dx} + \frac{y}{x} = 3e^{2x}, \qquad y\left(\frac{1}{2}\right) = 0.$$

Solution 9.15

Comparing terms with the general form of equation (9.18) we see that $k(x) = 3e^{2x}$ and $l(x) = \frac{1}{x}$. The integrating factor $p(x)$ is given by

$$p(x) = e^{\int l(x)dx} = e^{\int \frac{1}{x}\,dx} = e^{\ln x} = x\,.$$

The differential equation can now be written in the form:

$$\frac{d}{dx}(xy) = 3xe^{2x},$$

and integrating gives

$$xy = \int 3xe^{2x}\,dx.$$

The right hand side is evaluated using integration by parts to give

$$xy = \frac{3}{2}xe^{2x} - \frac{3}{4}e^{2x} + C,$$

and hence the general solution is

$$y = \frac{3}{2}e^{2x} - \frac{3}{4x}e^{2x} + \frac{C}{x}\,.$$

Substituting $y\left(\frac{1}{2}\right) = 0$ gives $C = 0$ and so the particular solution is

$$y = \frac{3}{2}e^{2x}\left(1 - \frac{1}{2x}\right)\,.$$

Problem 9.16

Find the particular solution which satisfies the following differential equation and initial condition, for $\pi/4 \leq x < \pi$

$$\sin x \frac{dy}{dx} + 2y \cos x + 2\sin^2 x \cos x = 0, \qquad y(\pi/4) = 1.$$

Solution 9.16

First rewrite the differential equation in the same form as equation (9.18). Divide through by the coefficient of $\frac{dy}{dx}$, which is $\sin x$, and take the last term over to the right hand side to obtain

$$\frac{dy}{dx} + 2y\frac{\cos x}{\sin x} = -2\sin x \cos x.$$

Note that $\sin x$ is never zero over the given interval for x.
The integrating factor $p(x)$ is given by

$$p(x) = \mathrm{e}^{\int l(x)dx},$$

and

$$2\int \frac{\cos x}{\sin x}\, dx = 2\ln(\sin x) = \ln(\sin^2 x).$$

Therefore

$$p(x) = \mathrm{e}^{\ln(\sin^2 x)} = \sin^2 x \ .$$

Using this integrating factor the differential equation becomes

$$\begin{aligned} \frac{d}{dx}\left(y\sin^2 x\right) &= -2\sin^3 x \cos x, \\ y\sin^2 x &= \int -2\sin^3 x \cos x \;\; dx \ . \end{aligned}$$

Carrying out the integration we have

$$y\sin^2 x = -\frac{1}{2}\sin^4 x + C,$$

and hence the general solution is

$$y = -\frac{1}{2}\sin^2 x + \frac{C}{\sin^2 x} \ .$$

Substituting $y(\pi/4) = 1$ gives $C = \frac{5}{8}$ and the particular solution is

$$y = -\frac{1}{2}\sin^2 x + \frac{5}{8\sin^2 x} \ .$$

Summary

In this section we have solved linear first order differential equations using integrating factors. This is the most general of the methods described so far and often requires greater effort than the earlier methods.
For linear differential equations of the type

$$\frac{dy}{dx} + l(x)y = k(x),$$

the general solution is found using

$$p(x)y = \int p(x)k(x)\ dx,$$

where $p(x)$ is the integrating factor

$$p(x) = \mathrm{e}^{\int l(x)dx}\ .$$

This concludes the description of the most common methods of solving first order differential equations. The sequence in which they have been described reflects the increasing level of difficulty associated with finding analytical solutions. Some differential equations can be solved by more than one method. The only difference will be in the effort expended.

EXERCISES

Find the particular solutions to the following first order linear differential equations by first obtaining the appropriate integrating factor.

9.35. $\dfrac{dy}{dx} + y = \mathrm{e}^{-x}, \qquad y(0) = 4$

9.36. $\dfrac{dy}{dx} + y = 2 + 2x, \qquad y(0) = 2$

9.37. $\dfrac{dx}{dt} + (\frac{1}{t} + 1)x = 3t, \qquad x(0) = 6$

9.38. $\dfrac{di}{dt} - 6i = 10 \sin 2t, \qquad i(0) = 0\ .$

9.3 Second order equations

9.3.1 Linear equations with constant coefficients

We only consider *linear equations* with *constant coefficients* of the form

$$a\frac{d^2y}{dx^2} + b\frac{dy}{dx} + cy = f(x), \tag{9.24}$$

where a, b and c are real numbers and $a \neq 0$.

Equation (9.24) can be subdivided into two categories, homogenous equations, where $f(x) = 0$ and non-homogenous equations.

The homogenous equation is always solved to find a *complementary function.* If $f(x) \neq 0$, an additional function, called a *particular integral,* is added to the complementary function to give the general solution. For a homogenous differential equation the complementary function *is* the general solution.

First we note some features regarding the solutions of differential equations. The functions y_1, y_2, ...,y_n are *linearly independent* if the *only* solution of

$$C_1y_1 + C_2y_2 + \ldots + C_ny_n = 0$$

is $C_1 = C_2 = \ldots = C_n = 0$. Otherwise the functions are *linearly dependent.* For example, the functions $y_1(x) = \sin x$ and $y_2(x) = x$ are linearly independent, since the only values of C_1 and C_2 for which

$$C_1 \sin x + C_2x = 0,$$

for *all* x are $C_1 = C_2 = 0$. On the other hand the functions $y_1(x) = x$ and $y_2(x) = 3x$ are linearly dependent since $C_1 = 3$ and $C_2 = -1$ are solutions of

$$C_1x + C_2(3x) = 0\ .$$

Two functions are linearly dependent if and only if one is a constant multiple of the other. If $y_1(x) \neq 0$ and $y_2(x) \neq Cy_1(x)$ then y_1 and y_2 are linearly independent. The following theorem indicates the importance of linear independence in constructing the general solution of a homogenous differential equation.

Theorem 9.1

If $y_1(x)$ and $y_2(x)$ are linearly independent solutions of

$$a(x)\frac{d^2y}{dx^2} + b(x)\frac{dy}{dx} + c(x)y = 0$$

then the general solution is

$$y(x) = C_1y_1(x) + C_2y_2(x)$$

where C_1 and C_2 are constants.

Proof

We shall only prove that if y_1 and y_2 are solutions of the differential equation then $y = C_1 y_1 + C_2 y_2$ is also a solution for any choice of C_1 and C_2. The proof that *every* solution is of that form is beyond the scope of this book.

If y_1 and y_2 are solutions then

$$a(x)\frac{d^2y_1}{dx^2} + b(x)\frac{dy_1}{dx} + c(x)y_1 = 0,$$
$$a(x)\frac{d^2y_2}{dx^2} + b(x)\frac{dy_2}{dx} + c(x)y_2 = 0\ .$$

Multiplying by C_1 and C_2 respectively and then adding gives

$$a(x)\left[C_1\frac{d^2y_1}{dx^2} + C_2\frac{d^2y_2}{dx^2}\right] + b(x)\left[C_1\frac{dy_1}{dx} + C_2\frac{dy_2}{dx}\right] + c(x)\left[C_1y_1 + C_2y_2\right] = 0.$$

Thus $y = C_1 y_1 + C_2 y_2$ is a solution for any choice of C_1 and C_2. Thus if we find two linearly independent solutions then the general solution is a linear combination of them. This is called the *principle of superposition.*□

9.3.2 Solution of the homogenous equation

The solution of a homogenous equation consists of exponentials whose indices are the roots of the *auxiliary equation* or *characteristic equation*

$$am^2 + bm + c = 0. \tag{9.25}$$

First of all we show that an exponential form does satisfy the homogenous version of equation (9.24). Suppose $y = A\mathrm{e}^{mx}$ where A is an arbitrary constant, then

$$\frac{dy}{dx} = Am\mathrm{e}^{mx} \quad \text{and} \quad \frac{d^2y}{dx^2} = Am^2\mathrm{e}^{mx}\ .$$

Substituting these derivatives into equation (9.24) we obtain

$$Aam^2\mathrm{e}^{mx} + Abm\mathrm{e}^{mx} + Ac\mathrm{e}^{mx} = A\mathrm{e}^{mx}\left(am^2 + bm + c\right) = 0\ .$$

So if m is a root of the auxiliary equation then $y = A\mathrm{e}^{mx}$ is a solution of the differential equation.

Equation (9.25) has two roots which can be found using

$$m = \frac{-b \pm \sqrt{b^2 - 4ac}}{2a}\ .$$

The roots fall into one of the following categories, depending on the value of the discriminant $\sqrt{b^2 - 4ac}$

– two distinct real roots if $b^2 - 4ac > 0$,
– two complex conjugate roots if $b^2 - 4ac < 0$,
– repeated real root if $b^2 - 4ac = 0$.

If there are two distinct roots (real or complex), m_1 and m_2 say, then the complementary function and hence the general solution is given by

$$y = Ae^{m_1 x} + Be^{m_2 x} \; . \tag{9.26}$$

If there is a repeated root, m_1, we cannot write the general solution as

$$y = Ae^{m_1 x} + Be^{m_1 x},$$

because these are not linearly independent solutions.

A general rule is to multiply the exponential by the independent variable, i.e. if $y = xe^{m_1 x}$ then

$$\frac{dy}{dx} = e^{m_1 x} + m_1 x e^{m_1 x}, \qquad \frac{d^2y}{dx^2} = 2m_1 e^{m_1 x} + m_1^2 x e^{m_1 x} \; .$$

Substituting into equation (9.24) and simplifying gives

$$xe^{m_1 x}\left(am_1^2 + bm_1 + c\right) + e^{mx}\left(2am_1 + b\right) = 0 \; .$$

The first bracketed term is zero because m_1 satisfies the auxiliary equation. The second bracketed term is also zero because, for repeated roots, the discriminant is zero giving $m_1 = -\frac{b}{2a}$, hence $2am_1 + b = 0$.
We have shown that $xe^{m_1 x}$ satisfies the differential equation and is linearly independent of $e^{m_1 x}$. Thus for a repeated root the general solution is

$$y = (A + Bx)e^{m_1 x} \; . \tag{9.27}$$

Note that there are two arbitrary constants. For a general n^{th} order differential equation there will be n arbitrary constants.

Problem 9.17

Find the general solution of the following differential equation

$$\frac{d^2y}{dx^2} + 3\frac{dy}{dx} + 2y = 0 \; . \tag{9.28}$$

Solution 9.17

Equation (9.28) is linear and homogenous and therefore the general solution is defined by finding the roots of the associated auxiliary equation

$$m^2 + 3m + 2 = 0.$$

The roots are $m = -2$ and $m = -1$ and so the general solution is

$$y = A\mathrm{e}^{-2x} + B\mathrm{e}^{-x} \ .$$

Problem 9.18

Find the general solution of the following differential equation

$$\frac{d^2y}{dx^2} + 2\frac{dy}{dx} + y = 0 \ . \tag{9.29}$$

Solution 9.18

The associated auxiliary equation is

$$m^2 + 2m + 1 = 0.$$

This has the repeated root $m = -1$ and so the general solution is

$$y = (A + Bx)\,\mathrm{e}^{-x} \ .$$

To determine the arbitrary constants in the general solution requires two initial conditions. Usually, initial values of y and $\frac{dy}{dx}$ are specified but sometimes values of y for different x values is given. This is called a boundary value problem.

Let us now find particular solutions for problems 9.17 and 9.18 satisfying the initial conditions

$$y(0) = 2, \qquad \frac{dy}{dx}(0) = 0 \ .$$

The first condition is easily substituted but we must differentiate the general solution before we can substitute the derivative condition.

Solution 9.17

The general solution was found to be

$$\begin{aligned} y &= A\mathrm{e}^{-2x} + B\mathrm{e}^{-x}; \\ \text{therefore} \qquad \frac{dy}{dx} &= -2A\mathrm{e}^{-2x} - B\mathrm{e}^{-x} \ . \end{aligned}$$

Substituting $y(0) = 2$ and $\frac{dy}{dx}(0) = 0$ gives $2 = A + B$, $0 = -2A - B$. These give $A = -2$, $B = 4$ and the particular solution is

$$y = -2\mathrm{e}^{-2x} + 4\mathrm{e}^{-x} \ .$$

Solution 9.18

The general solution was found to be

$$\begin{aligned} y &= (A+Bx)\,\mathrm{e}^{-x}; \\ \text{therefore} \qquad \frac{dy}{dx} &= B\mathrm{e}^{-x} - (A+Bx)\,\mathrm{e}^{-x}\ . \end{aligned}$$

Substituting $y(0) = 2$ and $\frac{dy}{dx}(0) = 0$ gives $2 = A$, $0 = B - A$. Thus $A = 2$, $B = 2$ and the particular solution is

$$y = 2\,(1+x)\,\mathrm{e}^{-x}\ .$$

9.3.3 Simple harmonic motion

Referring to equation (9.24), the special case where $a = 1, b = 0$, $c > 0$ and $f(x) = 0$ is worthy of further investigation. Let $c = \omega^2$ so that

$$\frac{d^2y}{dt^2} + \omega^2 y = 0\ . \tag{9.30}$$

The auxiliary equation

$$m^2 + \omega^2 = 0$$

has complex roots given by $m = \pm\,\omega\mathrm{i}$ and the general solution is therefore

$$y = A\mathrm{e}^{\mathrm{i}\omega t} + B\mathrm{e}^{-\mathrm{i}\omega t}\ .$$

This is not the most convenient form but if we recall the equivalence of polar and exponential forms from chapter 3 we can rewrite these exponentials

$$\mathrm{e}^{\mathrm{i}\omega t} = \cos\omega t + \mathrm{i}\sin\omega t, \qquad \mathrm{e}^{-\mathrm{i}\omega t} = \cos\omega t - \mathrm{i}\sin\omega t,$$

and hence

$$y = A\mathrm{e}^{\mathrm{i}\omega t} + B\mathrm{e}^{-\mathrm{i}\omega t} = C\sin\omega t + D\cos\omega t,$$

where $C = A + B$ and $D = \mathrm{i}(A - B)$.

If $y = C\sin\omega t + D\cos\omega t$, then

$$\begin{aligned} \frac{dy}{dt} &= \omega C\cos\omega t - \omega D\sin\omega t, \\ \frac{d^2y}{dt^2} &= -\omega^2 C\sin\omega t - \omega^2 D\cos\omega t = -\omega^2 y \end{aligned}$$

and hence this form for y satisfies the differential equation.
The solution is oscillatory, a fact made more obvious if y is written as

$$y = R\cos(\omega t + \phi),$$

where $R = \sqrt{C^2 + D^2}$ is the amplitude of the oscillation, ω is the frequency of oscillation and ϕ is a lead or lag term depending on its sign. The important points are the constant frequency ω and the constant amplitude R. This type of motion is referred to as *simple harmonic motion* (SHM).

Problem 9.19

Find the solution to the following second order differential equation with given initial conditions

$$\frac{d^2y}{dt^2} + 36y = 0, \quad y(0) = 0, \quad \frac{dy}{dt}(0) = 3\,.$$

Solution 9.19

The equation is linear and homogenous and the auxiliary equation is

$$m^2 + 36 = 0.$$

The roots are $m = \pm 6\mathrm{i}$ and so the general solution is

$$y = C\cos 6t + D\sin 6t\ .$$

Substituting the initial conditions gives $0 = C$ and $3 = 6D$, so

$$y = \frac{1}{2}\sin 6t,$$

which can be verified by substitution in the differential equation.

Problem 9.20

The motion of the bungee jumper as described in the introduction is an example of SHM. The motion of the jumper is given by

$$m\frac{d^2y}{dt^2} = -ky\ .$$

Here $k > 0$ represents the stiffness of the cord, m is the mass of the jumper and y is the distance from the equilibrium position, the position where the jumper would hang stationary on a stretched cord. Solve the differential equation subject to the initial conditions $y(0) = y_0$, the height of the bridge above the equilibrium position, and $\frac{dy}{dt}(0) = v_0$, the speed at which the jumper leaves the bridge. What is the solution if the jumper is initially at rest, $v(0) = 0$?

Solution 9.20

The differential equation can be written in the standard form of SHM:

$$\frac{d^2y}{dt^2} + \frac{k}{m}y = 0\ .$$

This corresponds to $\omega = \sqrt{\frac{k}{m}}$ and so the frequency of the oscillation is dependent on the stiffness of the cord and the mass of the jumper. A very stiff cord (large k) and a light jumper (small m) will give a high frequency of oscillation, whereas a heavy jumper (large m) on a stretchy cord (small k) will have a low frequency of oscillation.

The general solution of the differential equation is then

$$y(t) = A\cos\left(\sqrt{\frac{k}{m}}t\right) + B\sin\left(\sqrt{\frac{k}{m}}t\right)\ .$$

The values of A and B are found by substituting the initial conditions

$$\begin{aligned} y(t) &= A\cos\left(\sqrt{\frac{k}{m}}t\right) + B\sin\left(\sqrt{\frac{k}{m}}t\right), \\ \frac{dy}{dt}(t) &= -A\sqrt{\frac{k}{m}}\sin\left(\sqrt{\frac{k}{m}}t\right) + B\sqrt{\frac{k}{m}}\cos\left(\sqrt{\frac{k}{m}}t\right)\ . \end{aligned}$$

We have $y_0 = A$ and $v_0 = B\sqrt{\frac{k}{m}}$, therefore the particular solution is

$$y(t) = y_0\cos\left(\sqrt{\frac{k}{m}}t\right) + v_0\sqrt{\frac{m}{k}}\cos\left(\sqrt{\frac{k}{m}}t\right)\ .$$

This solution simplifies if the jumper has no initial speed, i.e. $v_0 = 0$, to

$$y(t) = y_0\cos\left(\sqrt{\frac{k}{m}}t\right)\ .$$

This solution is oscillatory because of the cosine function and therefore could possibly simulate the bouncing motion of the jumper. However, it predicts the motion of the jumper has a constant amplitude y_0 and will oscillate for ever!

The next section develops a more realistic model of the motion by including resistance terms which give a solution that is damped, i.e. an amplitude that decays with time.

9.3.4 Damped SHM

We now consider damped SHM where the roots of the auxiliary equation have non-zero real and imaginary parts. In general, the sum of two exponentials whose indices are complex conjugates can be expressed as

$$Ae^{(\alpha+i\beta)t} + Be^{(\alpha-i\beta)t} = e^{\alpha t}(C\sin\beta t + D\cos\beta t)\ .$$

The bracketed term describes SHM but the amplitude is now $R = e^{\alpha t}\sqrt{C^2 + D^2}$ which is *varying*. If α is positive then the amplitude grows exponentially, if α is negative the amplitude decays exponentially and is called *damped* SHM.

Problem 9.21

Find the general solution of the following differential equation

$$\frac{d^2y}{dx^2} + 4\frac{dy}{dx} + 5y = 0\ . \tag{9.31}$$

Solution 9.21

Equation (9.31) is linear and homogenous and therefore the general solution is defined by finding the roots of the associated auxiliary equation

$$m^2 + 4m + 5 = 0.$$

The roots of this quadratic are complex and are found using the standard equation for the roots of a quadratic

$$m = \frac{-b \pm \sqrt{b^2 - 4ac}}{2a} = \frac{-4 \pm \sqrt{16 - 20}}{2} = -2 \pm i\ .$$

The general solution is

$$y = e^{-2x}\left(Ae^{ix} + Be^{-ix}\right) = e^{-2x}\left(C\sin x + D\cos x\right)\ .$$

The term in brackets is that of SHM but the e^{-2x} represents a decaying amplitude and the solution displays *damped* SHM.

Problem 9.22

In an electrical circuit the charge, q, on a condenser satisfies the equation

$$\frac{d^2q}{dt^2} + 2\frac{dq}{dt} + 2q = 0. \tag{9.32}$$

At $t = 0$, the value of q is Q and $\frac{dq}{dt} = 0$. Express q in terms of t.

Solution 9.22

First solve the auxiliary equation

$$m^2 + 2m + 2 = 0.$$

The roots are

$$m = \frac{-2 \pm \sqrt{4-8}}{2} = -1 \pm \mathrm{i}\,.$$

The general solution is

$$q = A\mathrm{e}^{(-1+\mathrm{i})t} + B\mathrm{e}^{(-1-\mathrm{i})t} = \mathrm{e}^{-t}\left(C\sin t + D\cos t\right)\ .$$

We now use the initial conditions to find the values of C and D. To use the second condition we must differentiate the general solution.

$$\frac{dq}{dt} = \mathrm{e}^{-t}\left(C\cos t - D\sin t\right) - \mathrm{e}^{-t}\left(C\sin t + D\cos t\right)\ .$$

Substituting the initial conditions $q(0) = Q$ and $\frac{dq}{dt}(0) = 0$ gives

$$Q = D, \qquad\qquad 0 = C - D$$

Therefore $C = D = Q$ and the solution is

$$q = Q\mathrm{e}^{-t}\left(\sin t + \cos t\right)\quad .$$

As in Problem 9.19, the solution is the product of two parts. The sine and cosine terms represent oscillatory behaviour with a constant frequency of 1 hertz, while the exponential term represents a decaying amplitude.

Summary

The solution to the homogenous second order linear differential equation with constant coefficients

$$a\frac{d^2y}{dx^2} + b\frac{dy}{dx} + cy = 0,$$

is determined by the roots of the auxiliary equation

$$am^2 + bm + c = 0.$$

The form of the solution depends on the value of the discriminant $\sqrt{b^2 - 4ac}$:

- two distinct real roots if $b^2 - 4ac > 0$,
- two conjugate complex roots if $b^2 - 4ac < 0$,
- repeated real root if $b^2 - 4ac = 0$.

If there are two distinct roots, m_1 and m_2, the general solution is

$$y = A\mathrm{e}^{m_1 x} + B\mathrm{e}^{m_2 x} .$$

For a repeated root, m_1, the general solution is

$$y = (A + Bx)\mathrm{e}^{m_1 x} .$$

If the distinct roots are complex, $\alpha \pm i\beta$, then the general solution is

$$y = \mathrm{e}^{\alpha x}\left(C \sin \beta x + D \cos \beta x\right) .$$

EXERCISES

For the following differential equations write down the characteristic equation, find the roots and hence write down the general solution.

9.39. $\dfrac{d^2y}{dx^2} - 2\dfrac{dy}{dx} - 15y = 0$

9.40. $\dfrac{d^2y}{dx^2} = 4y$

9.41. $\dfrac{d^2y}{dx^2} - 2\dfrac{dy}{dx} + y = 0$

9.42. $\dfrac{d^2y}{dx^2} - 6\dfrac{dy}{dx} + 10y = 0$

9.43. $\dfrac{d^2y}{dx^2} + 36y = 0$

9.44. $\dfrac{d^2y}{dx^2} - 3\dfrac{dy}{dx} + 2y = 0$

9.45. The equation of motion of a body oscillating on a spring is

$$\frac{d^2x}{dt^2} + 100x = 0$$

where x is the displacement in metres of the body from its equilibrium position after time t seconds. Express x in terms of t given that at time $t = 0$, $x = 2$ m and $\frac{dx}{dt} = 0$ m/s.

9.46. The oscillations of a heavily damped pendulum satisfy

$$\frac{d^2x}{dt^2} + 7\frac{dx}{dt} + 12x = 0,$$

where x cm is the horizontal displacement, measured from the vertical equilibrium position, of the bob on the end of the pendulum at time t seconds. The initial displacement is equal to 3 cm and the initial velocity, $\frac{dx}{dt}$, is 6 cm/s. Solve the equation for x.

9.47. The differential equation

$$\frac{d^2i}{dt^2} + \frac{R}{L}\frac{di}{dt} + \frac{1}{LC}i = 0,$$

represents a current i flowing in an electrical circuit containing a resistance R, an inductance L and a capacitance C connected in series. The initial conditions at time $t = 0$ are $i = 0$ and $\frac{di}{dt} = 100$. If $L = 0.20$ henrys and $C = 80 \times 10^{-6}$ farads, find the solutions that correspond to (a) $R = 100$ ohms and (b) $R = 223.6$ ohms.

9.48. The oscillations of a critically damped pendulum satisfy the equation

$$\frac{d^2x}{dt^2} + 6\frac{dx}{dt} + 9x = 0,$$

where x cm is the horizontal displacement, measured from the vertical equilibrium position, of the bob on the end of the pendulum at time t seconds. The initial displacement is equal to 3 cm and the initial velocity is 6 cm/sec directed away from the equilibrium position. Find x as a function of t.

9.49. In a galvanometer the deflection θ satisfies the differential equation

$$\frac{d^2\theta}{dt^2} + 2\frac{d\theta}{dt} + 10\theta = 0$$

Solve the equation for θ given the initial values $\theta = 0$ and $\frac{d\theta}{dt} = 0.3$.

9.3.5 Particular integrals involving elementary functions

Let us now return to the non-homogenous differential equation (9.24). The general solution consists of two parts, the complementary function and the particular integral. The complementary function is the general solution of the homogenous equation and the particular integral is associated with the function $f(x)$ on the right hand side.

The particular integral is usually found by substituting the general form of $f(x)$ into the differential equation and then comparing coefficients. This is called the method of undetermined coefficients and is best illustrated by means of an example.

Problem 9.23

Find the general solution of the differential equation

$$\frac{d^2y}{dx^2} + 3\frac{dy}{dx} + 2y = x^2 - 5x + 2 \ . \tag{9.33}$$

Solution 9.23

The complementary function, which we denote by y_C to distinguish it from the general solution, is defined by the roots of the auxiliary equation

$$m^2 + 3m + 2 = 0.$$

The roots are $m = -2$ and $m = -1$, so the complementary function is

$$y_C = A\mathrm{e}^{-2x} + B\mathrm{e}^{-x}.$$

The particular integral, y_P, is found by substituting the general form of the right hand side, $f(x)$. Here $f(x)$ is a quadratic in x, so we try

$$y_P = px^2 + qx + r\ .$$

Substituting this into equation (9.33) gives

$$\begin{aligned} 2p + 3(2px + q) + 2(px^2 + qx + r) &= x^2 - 5x + 2 \\ 2px^2 + (6p + 2q)x + (2p + 3q + 2r) &= x^2 - 5x + 2. \end{aligned}$$

Comparing coefficients we find values of p, q and r which satisfy

$$2p = 1 \qquad 6p + 2q = -5 \qquad 2p + 3q + r = 2$$

i.e. $p = \frac{1}{2}$, $q = -4$ and $r = \frac{7}{8}$. Therefore the particular integral is

$$y_P = \frac{1}{2}x^2 - 4x + \frac{7}{8}$$

and the general solution is given by $y = y_C + y_P$

$$y = A\mathrm{e}^{-2x} + B\mathrm{e}^{-x} + \frac{1}{2}x^2 - 4x + \frac{7}{8}.$$

Provided you use a trial function which is the general form of $f(x)$, there should be few problems. The following table lists some common trial functions.

f(x)	**trial function**
polynomial of degree n	general polynomial of degree n
e^{kx}	$C\mathrm{e}^{kx}$
$\cos(kx)$ and/or $\sin(kx)$	$C\cos(kx) + D\sin(kx)$
$\mathrm{e}^{kx}\cos(qx)$ and/or $\mathrm{e}^{kx}\sin(qx)$	$\mathrm{e}^{kx}(C\cos(qx) + D\sin(qx))$

Note that the trial function contains both sines *and* cosines even when $f(x)$ only has one or the other.

Problem 9.24

Find the general solution of the differential equation

$$2\frac{d^2y}{dx^2} - 5\frac{dy}{dx} - 3y = 221\sin(2x)\ . \tag{9.34}$$

Solution 9.24

The complementary function is defined by finding the roots of the auxiliary equation

$$2m^2 - 5m - 3 = 0.$$

The roots are $m = -\frac{1}{2}$ and $m = 3$ and so the complementary function is

$$y_C = A\mathrm{e}^{-\frac{1}{2}x} + B\mathrm{e}^{3x}.$$

To find the particular integral we try the general form of the right hand side of equation (9.34),

$$y_P = C\sin 2x + D\cos 2x\ .$$

Substituting this expression into equation (9.34) gives

$$(-10C - 11D)\cos 2x + (-11C + 10D)\sin 2x = 221\sin 2x\ .$$

For this equation to be true for *all* values of x, the coefficients on the sine and cosine terms must be the same on both sides of the equation, i.e.

$$-10C - 11D = 0 \quad \text{and} \quad -11C + 10D = 221\ .$$

Thus $D = 10$ and $C = -11$. The particular integral is

$$y_P = -11\sin 2x + 10\cos 2x,$$

and the general solution is

$$y = A\mathrm{e}^{-\frac{1}{2}x} + B\mathrm{e}^{3x} - 11\sin 2x + 10\cos 2x\ .$$

A complication that regularly arises is that one or more of the terms in the complementary function also appears in $f(x)$. We then follow the same procedure for a repeated root in the auxiliary equation and multiply by the independent variable to get another linearly independent solution.

Problem 9.25

Find the general solution of the differential equation

$$\frac{d^2x}{dt^2} + 3\frac{dx}{dt} - 18x = 9\mathrm{e}^{3t}\ . \tag{9.35}$$

Solution 9.25

The auxiliary equation is

$$m^2 + 3m - 18 = 0,$$

which has roots $m = -6$ and $m = 3$. Therefore

$$x_C = Ae^{-6t} + Be^{3t}.$$

If we were to try a particular integral of the form

$$x_P = Ce^{3t} .$$

then it would simply be absorbed with the same term in x_C. So we try

$$x_P = Cte^{3t}$$

Substituting this expression for x_P into equation (9.35) gives

$$Ce^{3t}(6 + 9t) + 3Ce^{3t}(1 + 3t) - 18Cte^{3t} = 9e^{3t}$$

$$9Ce^{3t} = 9e^{3t} .$$

Hence $C = 1$, and the particular integral is $x_P = te^{3t}$.
The general solution is

$$x = Ae^{-6t} + Be^{3t} + te^{3t} .$$

Summary

This section has shown how to find a particular integral when the differential equation is not homogenous. The principle is straightforward; substitute the general form of the function on the right-hand side and then compare coefficients to find the constants.

If a term on the right hand-side appears in the complementary function then multiply the general term by the independent variable.

EXERCISES

Find the general solutions to the following homogenous second order differential equations.

9.50. $\dfrac{d^2y}{dx^2} - 4\dfrac{dy}{dx} + 5y = 3e^{2x}$

9.51. $\dfrac{d^2y}{dx^2} + \dfrac{dy}{dx} - 6y = 3x + e^{3x}$

9.52. $\frac{d^2x}{dt^2} - 2\frac{dx}{dt} = 3\cos(2t)$

9.53. The equation of motion of a body oscillating on a spring is

$$\frac{d^2x}{dt^2} + 6\frac{dx}{dt} + 25x = P(t),$$

where x is the displacement in metres of the body from its equilibrium position after time t seconds and $P(t) = 5\cos t$ is an extra applied oscillatory force. Express x in terms of t given that at time $t = 0$, $x = 2$ m and $\frac{dx}{dt} = 0$ m/s.

9.54. The differential equation

$$L\frac{d^2q}{dt^2} + R\frac{dq}{dt} + \frac{1}{C}q = E(t),$$

represents the charge, q, on the capacitor in an electrical circuit containing a resistance R, an inductance L and a capacitance C connected in series with an impressed voltage $E(t) = 110\cos(120t)$. The initial conditions at time $t = 0$ are $q = 0$ and $\frac{dq}{dt} = 50$. If $L = 10$ henrys, $R = 3 \times 10^3$ ohms and $C = 0.4 \times 10^{-5}$ farads, express the charge as a function of time.

9.55. A couple reach retirement age with a regular pension of £750 per month and the total value of their house, savings and investments amounts to £180,000. The pension is sufficient to cover all their living expenses and pay for a three-week vacation each year, leaving the £180,000 of assets to accumulate interest for their children to inherit when the couple die.

However, this particular elderly couple want to 'live a little'. They reckon on surviving for another twenty years and want to maximise their spending power by realising their assets and gradually using up the capital. They calculate how much they can withdraw each month by modelling the cashflow as a continuous function rather than as discrete interest payments and withdrawals. The capital is assumed to generate r % interest continuously over each month.

A possible differential equation modelling the amount of capital, $p(t)$, is

$$\frac{dp}{dt} = rp - W \ .$$

On the right hand side the first term, rp, indicates an increase in capital due to accruing interest, while the term, $-W$, reflects the reduction in capital due to withdrawals.

Solve this differential equation subject to the initial condition $p(0) = P_0$. Find W if $r = 0.5\%, p_0 = £180,000$ and $p = 0$ after twenty years.

9.56. Aluminium and tinplate is used extensively in the production of the billions of cans required by the beverages industry each year. Over 100 billion aluminium cans during 1995 required in the USA alone. Apart from the sheer quantity of metal involved another important aspect is the stringent tolerance on thickness variations, currently around 0.005 mm. Achieving such tolerances requires good models of metal behaviour during the forming stages.
The force $s(\theta)$ can be modelled using differential equations such as

$$\frac{d}{d\theta}\left(\frac{s}{k}-\frac{\pi}{4}\right)=\frac{\pi R\theta}{2(h_0+R\theta^2)}\pm\frac{R}{h_0+R\theta^2}, \tag{9.36}$$

where R, h_0 and k are constants.
Integrate this equation with respect to θ.
The range for θ is $0\le\theta\le\alpha$ and $h_0+R\alpha^2=h_1$, where α and h_1 are constants. There are two solutions because of the '$\pm$' in the differential equation. Using the solution corresponding to the '$+\frac{R}{h_0+R\theta^2}$' term show that the particular solution, s^+, which satisfies the condition

$$s(\theta=0)=\frac{\pi k}{4},$$

is given by

$$s^+=\frac{\pi}{4}+\frac{\pi}{4}\ln\left(\frac{h_0+R\theta^2}{h_0}\right)+\sqrt{\frac{R}{h_0}}\tan^{-1}\sqrt{\frac{R}{h_0}}\theta\,. \tag{9.37}$$

Using the solution corresponding to the '$-\frac{R}{h_0+R\theta^2}$' term show that the particular solution, s^-, which satisfies the condition

$$s(\theta=\alpha)=\frac{\pi k}{4},$$

is given by

$$s^-=\frac{\pi}{4}+\frac{\pi}{4}\ln\left(\frac{h_0+R\theta^2}{h_1}\right)-\sqrt{\frac{R}{h_0}}\tan^{-1}\sqrt{\frac{R}{h_0}}\theta+\sqrt{\frac{R}{h_0}}\tan^{-1}\sqrt{\frac{R}{h_0}}\alpha \tag{9.38}$$

The values of s^+ and s^- are identical for one value of θ in the range $0<\theta<\alpha$. Equate the two solutions and show that this value, θ_n, is given by

$$\theta_n=\sqrt{\frac{h_0}{R}}\tan\left[\frac{\sqrt{\frac{R}{h_0}}\tan^{-1}\sqrt{\frac{R}{h_0}}\alpha-\frac{\pi}{4}\ln\left(\frac{h_1}{h_0}\right)}{2\sqrt{\frac{R}{h_0}}}\right]. \tag{9.39}$$

Solutions to Exercises

1.1 Calculate determinants

$$\Delta = \begin{vmatrix} 3 & -7 \\ 5 & 2 \end{vmatrix} = 41, \ \Delta_x = \begin{vmatrix} 47 & -7 \\ 10 & 2 \end{vmatrix} = 164, \ \Delta_y = \begin{vmatrix} 3 & 47 \\ 5 & 10 \end{vmatrix} = -205$$

Hence the solution is $x = \frac{\Delta_x}{\Delta} = \frac{164}{41} = 4 \quad y = \frac{\Delta_y}{\Delta} = \frac{-205}{41} = -5.$

1.2 Calculate determinants

$$\begin{aligned} \Delta &= \begin{vmatrix} 1.985 & -1.358 \\ 0.953 & -0.652 \end{vmatrix} = -0.000046, \ \Delta_y = \begin{vmatrix} 1.985 & 2.212 \\ 0.953 & 1.062 \end{vmatrix} = 0.000034 \\ \Delta_x &= \begin{vmatrix} 2.212 & -1.358 \\ 1.062 & -0.652 \end{vmatrix} = 0.000028 \end{aligned}$$

Hence the solution is

$$x = \frac{\Delta_x}{\Delta} = \frac{0.000028}{-0.000046} = -0.609 \qquad y = \frac{\Delta_y}{\Delta} = \frac{0.000034}{-0.000046} = -0.739$$

1.3 If we calculate the determinant Δ, we find it is zero in both cases:

$$\text{(a)}\Delta = 2 \times 6 - 3 \times 4 = 0 \qquad \text{(b)}\Delta = 1 \times (-6) - (-2) \times 3 = 0$$

Thus neither pair of equations has a unique solution. The explanations are: (a) these equations represent parallel lines and so never cross, hence no solution; (b) these equations are multiples of each other, i.e. they both represent the same line and hence intersect at an infinite number of points.

1.4 We first calculate the four determinants, expanding about the first row.

$$\begin{aligned} \Delta &= \begin{vmatrix} 1 & 3 & 1 \\ 2 & 1 & 4 \\ 3 & 1 & -2 \end{vmatrix} = 1\begin{vmatrix} 1 & 4 \\ 1 & -2 \end{vmatrix} - 3\begin{vmatrix} 2 & 4 \\ 3 & -2 \end{vmatrix} + 1\begin{vmatrix} 2 & 1 \\ 3 & 1 \end{vmatrix} \\ &= -6 + 48 - 1 = 41 \\ \Delta_x &= \begin{vmatrix} 3 & 3 & 1 \\ -1 & 1 & 4 \\ 6 & 1 & -2 \end{vmatrix} = 3\begin{vmatrix} 1 & 4 \\ 1 & -2 \end{vmatrix} - 3\begin{vmatrix} -1 & 4 \\ 6 & -2 \end{vmatrix} + 1\begin{vmatrix} -1 & 1 \\ 6 & 1 \end{vmatrix} \end{aligned}$$

$$
\begin{aligned}
&= -18+66-7=41\\
\Delta_y &= \begin{vmatrix} 1 & 3 & 1\\ 2 & -1 & 4\\ 3 & 6 & -2 \end{vmatrix} = 1\begin{vmatrix} -1 & 4\\ 6 & -2\end{vmatrix} - 3\begin{vmatrix} 2 & 4\\ 3 & -2\end{vmatrix} + 1\begin{vmatrix} 2 & -1\\ 3 & 6\end{vmatrix}\\
&= -22+48+15=41\\
\Delta_z &= \begin{vmatrix} 1 & 3 & 3\\ 2 & 1 & -1\\ 3 & 1 & 6 \end{vmatrix} = 1\begin{vmatrix} 1 & -1\\ 1 & 6\end{vmatrix} - 3\begin{vmatrix} 2 & -1\\ 3 & 6\end{vmatrix} + 3\begin{vmatrix} 2 & 1\\ 3 & 1\end{vmatrix}\\
&= 7-45-3=-41
\end{aligned}
$$

and hence the solution is $x = \frac{\Delta_x}{\Delta} = 1$, $y = \frac{\Delta_y}{\Delta} = 1$, $z = \frac{\Delta_z}{\Delta} = -1$.

1.5
$$
\begin{array}{rrl}
x+3y+z= & 3 & (1)\\
2x+y+4z= & -1 & (2)\\
3x+y-2z= & 6 & (3)
\end{array}
\quad
\begin{array}{c}
\\
(2)-\tfrac{2}{1}\times(1)\\
(3)-\tfrac{3}{1}\times(1)
\end{array}
\quad
\begin{array}{rrl}
x+3y+z= & 3 & (1)\\
-5y+2z= & -7 & (4)\\
-8y-5z= & -3 & (5)
\end{array}
$$

Eliminate y from (5) using $(5) - \frac{-8}{-5}\times(4)$

$$
\begin{array}{rrl}
x+3y+z= & 3 & (1)\\
-5y+2z= & -7 & (4)\\
-\frac{41}{5}z= & \frac{41}{5} & (6)
\end{array}
$$

Back substitution gives $z=-1$, $y=1$, $x=1$.

1.6 Rearrange as

$$
\begin{array}{rrl}
y+2x+4z= & 17 & (1)\\
-3y+3x-z= & 0 & (2)\\
x+z= & 5 & (3)
\end{array}
\quad
\begin{array}{c}
\\
(2)-\tfrac{-3}{1}\times(1)\\
\\
\end{array}
\quad
\begin{array}{rrl}
y+2x+4z= & 17 & (1)\\
9x+11z= & 51 & (4)\\
x+z= & 5 & (3)
\end{array}
$$

Eliminate x from (3) using $(3) - \frac{1}{9}\times(4)$

$$
\begin{array}{rrl}
y+2x+4z= & 17 & (1)\\
9x+11z= & 51 & (4)\\
\frac{-2}{9}z= & \frac{-6}{9} & (5)
\end{array}
$$

Back substitution gives $z=3$, $y=1$, $x=2$.

1.7 A has 2 rows and 3 columns and is therefore order 2×3. B has 3 rows and 3 columns and is order 3×3. C has 1 row and 3 columns and is order 1×3.
1.8 The first subscript gives the row number and the second subscript the column number. Therefore $a_{11}=3$, $b_{32}=6$ and $c_{13}=1$.
1.9 The products are: $a_{12}\times b_{33} = -2\times1 = -2$ $\quad a_{22}\times c_{12} = 0\times(-3) = 0$.
1.10 A is a square matrix therefore it is possible to calculate powers of A.

$$
\begin{aligned}
A^2 &= \begin{pmatrix} 1 & 2\\ 0 & 1\end{pmatrix}\begin{pmatrix} 1 & 2\\ 0 & 1\end{pmatrix} = \begin{pmatrix} 1 & 4\\ 0 & 1\end{pmatrix}\\
A^3 &= \begin{pmatrix} 1 & 4\\ 0 & 1\end{pmatrix}\begin{pmatrix} 1 & 2\\ 0 & 1\end{pmatrix} = \begin{pmatrix} 1 & 6\\ 0 & 1\end{pmatrix}\\
A^n &= \begin{pmatrix} 1 & 2n\\ 0 & 1\end{pmatrix}
\end{aligned}
$$

1.11

$$\begin{pmatrix} 4 & -3 & 1 \\ 2 & 0 & 1 \\ 1 & 1 & -1 \end{pmatrix}\begin{pmatrix} 3 & -2 & 0 \\ -1 & 11 & 0 \\ 0 & 6 & 1 \end{pmatrix} = \begin{pmatrix} 15 & -35 & 1 \\ 6 & 2 & 1 \\ 2 & 3 & -1 \end{pmatrix}$$

1.12 $\det A = 1$, $\det B = 1$

$$\begin{aligned} AB &= \begin{pmatrix} 3 & 2 \\ 1 & 1 \end{pmatrix}\begin{pmatrix} 1 & -2 \\ -1 & 3 \end{pmatrix} = \begin{pmatrix} 1 & 0 \\ 0 & 1 \end{pmatrix} \\ BA &= \begin{pmatrix} 1 & -2 \\ -1 & 3 \end{pmatrix}\begin{pmatrix} 3 & 2 \\ 1 & 1 \end{pmatrix} = \begin{pmatrix} 1 & 0 \\ 0 & 1 \end{pmatrix} \end{aligned}$$

1.13 $\det B = -1(-2-0) - 0 + 2(-80+3) = -152$. The matrix of co-factors is

$$\begin{pmatrix} -2 & 20 & -77 \\ 16 & -8 & 8 \\ 2 & -20 & 1 \end{pmatrix}$$

$$\text{adj}B = \begin{pmatrix} -2 & 16 & 2 \\ 20 & -8 & -20 \\ -77 & 8 & 1 \end{pmatrix} \qquad B^{-1} = \frac{-1}{152}\begin{pmatrix} -2 & 16 & 2 \\ 20 & -8 & -20 \\ -77 & 8 & 1 \end{pmatrix}$$

1.14 Find the inverse: $\det A = 41$, $A^{-1} = \frac{1}{41}\begin{pmatrix} 9 & 5 \\ -1 & 4 \end{pmatrix}$. Solution is:

$$X = A^{-1}B = \begin{pmatrix} x \\ y \end{pmatrix} = \frac{1}{41}\begin{pmatrix} 9 & 5 \\ -1 & 4 \end{pmatrix}\begin{pmatrix} -23 \\ 66 \end{pmatrix} = \frac{1}{41}\begin{pmatrix} 123 \\ 287 \end{pmatrix} = \begin{pmatrix} 3 \\ 7 \end{pmatrix}$$

1.15 Find the inverse, $\det A = 800$, $A^{-1} = \frac{1}{800}\begin{pmatrix} 1 & -700 \\ -1 & 1500 \end{pmatrix}$. Solution is:

$$\begin{pmatrix} N \\ S \end{pmatrix} = \frac{1}{800}\begin{pmatrix} 1 & -700 \\ -1 & 1500 \end{pmatrix}\begin{pmatrix} 7200 \\ 8 \end{pmatrix} = \frac{1}{800}\begin{pmatrix} 1600 \\ 4800 \end{pmatrix} = \begin{pmatrix} 2 \\ 6 \end{pmatrix}$$

1.16 $\det A = 5 \times (-8) - 1 \times (-7) - 2 \times (1) = -35$.

Matrix of co-factors is $\begin{pmatrix} -8 & 7 & 1 \\ -16 & 14 & -33 \\ 3 & -7 & 4 \end{pmatrix}$

$$\text{adj}A = \begin{pmatrix} -8 & -16 & 3 \\ 7 & 14 & -7 \\ 1 & -33 & 4 \end{pmatrix} \qquad A^{-1} = \frac{-1}{35}\begin{pmatrix} -8 & -16 & 3 \\ 7 & 14 & -7 \\ 1 & -33 & 4 \end{pmatrix}$$

$$\begin{pmatrix} x \\ y \\ z \end{pmatrix} = \frac{-1}{35}\begin{pmatrix} -8 & -16 & 3 \\ 7 & 14 & -7 \\ 1 & -33 & 4 \end{pmatrix}\begin{pmatrix} -1 \\ -3 \\ -7 \end{pmatrix} = \frac{-1}{35}\begin{pmatrix} 35 \\ 0 \\ 70 \end{pmatrix} = \begin{pmatrix} -1 \\ 0 \\ -2 \end{pmatrix}$$

2.1 (a) $5\mathbf{i}-3\mathbf{j}-7\mathbf{k}$, (b) $3\mathbf{i}+6\mathbf{k}$, (c) $24\mathbf{i}+3\mathbf{j}+33\mathbf{k}$

2.2 (a) If $\mathbf{a}=-\frac{1}{\sqrt{2}}\mathbf{j}+\frac{1}{\sqrt{2}}\mathbf{k}$ then $|\mathbf{a}|=\sqrt{(-\frac{1}{2})^2+(\frac{1}{2})^2}=1$, $\mathbf{a}$ is a unit vector.

(b) Let $\mathbf{b}=-\frac{3}{5}\mathbf{i}-\frac{4}{5}\mathbf{j}$ then $|\mathbf{b}|=\sqrt{(-\frac{3}{5})^2+(\frac{-4}{5})^2}=1$, $\mathbf{b}$ is a unit vector.

2.3

$|\mathbf{a}|=\sqrt{3^2+(-1)^2+(-1)^2}=\sqrt{11}$, $\quad\hat{\mathbf{a}}=\frac{1}{\sqrt{11}}(3\mathbf{i}-\mathbf{j}-\mathbf{k})$

$|\mathbf{b}|=\sqrt{1^2+1^2+5^2}=3\sqrt{3}$, $\quad\hat{\mathbf{b}}=\frac{1}{3\sqrt{3}}(\mathbf{i}+\mathbf{j}+5\mathbf{k})$

$|\mathbf{c}|=\sqrt{(-1)^2+0^2+2^2}=\sqrt{5}$, $\quad\hat{\mathbf{c}}=\frac{1}{\sqrt{5}}(-\mathbf{i}+2\mathbf{k})$

2.4 Using components:

$$\mathbf{a}.\mathbf{b}=1\times1+1\times(-1)+1\times0=0\ .$$

As $\mathbf{a}$ and $\mathbf{b}$ are non-zero vectors, $\mathbf{a}$ and $\mathbf{b}$ are perpendicular and the angle between them is 90° or $\frac{\pi}{2}$ radians.

2.5 The scalar product is zero if the two vectors are perpendicular.

$$(a\mathbf{i}-2\mathbf{j}+\mathbf{k}).(2a\mathbf{i}+a\mathbf{j}-4\mathbf{k})=2a^2-2a-4=2(a-2)(a+1)=0\ .$$

Either $a=2$ or $a=-1$.

2.6 Let the three angles of the triangle be α, β and γ and α be the angle between $\mathbf{a}$ and $\mathbf{b}$. The two vectors subtending one of the other angles, say β, are $-\mathbf{a}$ and $\mathbf{b}-\mathbf{a}$ while the other angle, γ, is subtended by the vectors $-\mathbf{b}$ and $\mathbf{a}-\mathbf{b}$. Note the opposite signs on the third vector depending on which corner of the triangle is under consideration.

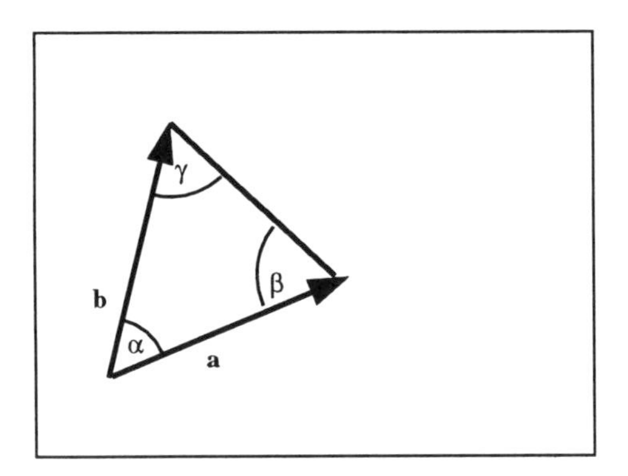

Fig. 10.1. Triangle in Exercise 2.6

The vectors involved are

$$\begin{aligned}\mathbf{a}&=2\mathbf{i}-\mathbf{j}+3\mathbf{k} &\quad \mathbf{a}-\mathbf{b}=\mathbf{i}-3\mathbf{j}-\mathbf{k}\\ \mathbf{b}&=\mathbf{i}+2\mathbf{j}+4\mathbf{k} &\quad \mathbf{b}-\mathbf{a}=-\mathbf{i}+3\mathbf{j}+\mathbf{k}\end{aligned}$$

The scalar product then gives $\mathbf{a}.\mathbf{b}=ab\cos\alpha$ and

$$(-\mathbf{a}).(\mathbf{b}-\mathbf{a}) = a|\mathbf{b}-\mathbf{a}|\cos\beta \qquad (-\mathbf{b}).(\mathbf{a}-\mathbf{b}) = b|\mathbf{a}-\mathbf{b}|\cos\gamma$$

and hence the angles are found using

$$\begin{aligned}
\cos\alpha &= \frac{\mathbf{a}.\mathbf{b}}{ab} = \frac{2-2+12}{\sqrt{14}\sqrt{21}} = 0.6999 \rightarrow \alpha = 45.6^o \\
\cos\beta &= \frac{(-\mathbf{a}).(\mathbf{b}-\mathbf{a})}{a|\mathbf{b}-\mathbf{a}|} = \frac{2+3-3}{\sqrt{14}\sqrt{11}} = 0.1612 \rightarrow \beta = 80.7^o \\
\cos\gamma &= \frac{(-\mathbf{b}).(\mathbf{a}-\mathbf{b})}{b|\mathbf{a}-\mathbf{b}|} = \frac{-1+6+4}{\sqrt{21}\sqrt{11}} = 0.5922 \rightarrow \gamma = 53.7^o
\end{aligned}$$

Summing the three angles: $\alpha+\beta+\gamma = 180^o$.

2.7 The displacement vector is $\mathbf{AB} = 4\mathbf{i}+3\mathbf{j}-7\mathbf{k}$. A unit vector in the direction of the force is $(3\mathbf{i}-\mathbf{j}+\mathbf{k})/\sqrt{11}$. Therefore $\mathbf{F} = 5(3\mathbf{i}-\mathbf{j}+\mathbf{k})/\sqrt{11}$.
Finally, work done $= \mathbf{F}.\mathbf{AB} = 3.02$ joules

2.8 Evaluate the determinant

$$\mathbf{a}\times\mathbf{b} = \begin{vmatrix} \mathbf{i} & \mathbf{j} & \mathbf{k} \\ 1 & 1 & -1 \\ 2 & -1 & 3 \end{vmatrix} = 2\mathbf{i}-5\mathbf{j}-3\mathbf{k}$$

2.9 The point D is $(-\frac{1}{2},-\frac{1}{2},2)$. Define the two vectors in component form

$$\mathbf{AB} = \mathbf{i}+2\mathbf{j}, \qquad \mathbf{BD} = -\frac{3}{2}\mathbf{i}-\frac{7}{2}\mathbf{j}+\mathbf{k}\,.$$

Evaluate the determinant

$$\mathbf{AB}\times\mathbf{BD} = \begin{vmatrix} \mathbf{i} & \mathbf{j} & \mathbf{k} \\ 1 & 2 & 0 \\ -\frac{3}{2} & -\frac{7}{2} & 1 \end{vmatrix} = 2\mathbf{i}-\mathbf{j}-\frac{1}{2}\mathbf{k}$$

To find a unit vector in the same direction we divide a vector by its magnitude $\sqrt{4+1+\frac{1}{4}} = \frac{\sqrt{21}}{2}$, hence a unit vector is $\frac{1}{\sqrt{21}}(4\mathbf{i}-2\mathbf{j}-\mathbf{k})$.

2.10 $\mathbf{r} = -\mathbf{i}-\mathbf{k}$. Hence $\mathbf{M} = \mathbf{r}\times\mathbf{F} = \begin{vmatrix} \mathbf{i} & \mathbf{j} & \mathbf{k} \\ -1 & 0 & -1 \\ 3 & 2 & -4 \end{vmatrix} = 2\mathbf{i}-7\mathbf{j}-2\mathbf{k}$.

2.11 Substitute the components into the third order determinant.

$$\mathbf{a}.(\mathbf{b}\times\mathbf{c}) = \begin{vmatrix} a_1 & a_2 & a_3 \\ b_1 & b_2 & b_3 \\ c_1 & c_2 & c_3 \end{vmatrix} = \begin{vmatrix} 2 & -3 & 1 \\ 3 & 1 & 2 \\ 1 & 4 & -2 \end{vmatrix} = -20-24+11 = -33$$

2.12 Evaluate both sides separately then compare.

$$\mathbf{b}\times\mathbf{c} = \begin{vmatrix} \mathbf{i} & \mathbf{j} & \mathbf{k} \\ b_1 & b_2 & b_3 \\ c_1 & c_2 & c_3 \end{vmatrix} = \mathbf{i}(b_2c_3-c_2b_3)-\mathbf{j}(b_1c_3-c_1b_3)+\mathbf{k}(b_1c_2-c_1b_2)$$

and so

$$\begin{aligned}\mathbf{a}\times(\mathbf{b}\times\mathbf{c}) &= \begin{vmatrix} \mathbf{i} & \mathbf{j} & \mathbf{k} \\ a_1 & a_2 & a_3 \\ (b_2c_3 - c_2b_3) & (b_1c_3 - c_1b_3) & (b_1c_2 - c_1b_2) \end{vmatrix} \\ &= \mathbf{i}[b_1(a_2c_2 + a_3c_3) - c_1(a_2b_2 + a_3b_3)] + \mathbf{j}[b_2(a_1c_1 + a_3c_3) \\ &\quad -c_2(a_1b_1 + a_3b_3)] + \mathbf{k}[b_3(a_1c_1 + a_2c_2) - c_3(a_1b_1 + a_2b_2)]\end{aligned}$$

Now find the terms on the right hand side of the identity.

$$\begin{aligned}(\mathbf{a}.\mathbf{c})\mathbf{b} - (\mathbf{a}.\mathbf{b})\mathbf{c} &= (a_1c_1 + a_2c_2 + a_3c_3)\mathbf{b} - (a_1b_1 + a_2b_2 + a_3b_3)\mathbf{c} \\ &= \mathbf{i}[b_1(a_1c_1 + a_2c_2 + a_3c_3) - c_1(a_1b_1 + a_2b_2 + a_3b_3)] \\ &\quad +\mathbf{j}[b_2(a_1c_1 + a_2c_2 + a_3c_3) - c_2(a_1b_1 + a_2b_2 + a_3b_3)] \\ &\quad +\mathbf{k}[b_3(a_1c_1 + a_2c_2 + a_3c_3) - c_3(a_1b_1 + a_2b_2 + a_3b_3)] \\ &= \mathbf{i}[b_1(a_2c_2 + a_3c_3) - c_1(a_2b_2 + a_3b_3)] \\ &\quad +\mathbf{j}[b_2(a_1c_1 + a_3c_3) - c_2(a_1b_1 + a_3b_3)] \\ &\quad +\mathbf{k}[b_3(a_1c_1 + a_2c_2) - c_3(a_1b_1 + a_2b_2)]\end{aligned}$$

The identity is proved.
2.13 Differentiate with respect to t.

$$\begin{aligned}\mathbf{v} &= \frac{d}{dt}(\mathbf{r}) = 2t\mathbf{i} - 4\cos(4t)\mathbf{j} + 2e^{2t}\mathbf{k} \\ \mathbf{a} &= \frac{d^2}{dt^2}(\mathbf{r}) = 2\mathbf{i} + 16\sin(4t)\mathbf{j} + 4e^{2t}\mathbf{k}\end{aligned}$$

Substituting the value $t = 0$ we find $\mathbf{v}(0) = -4\mathbf{j} + 2\mathbf{k}$, $\mathbf{a}(0) = 2\mathbf{i} + 4\mathbf{k}$.
2.14 (a) $|\mathbf{n}| = \sqrt{\cos^2\theta + \sin^2\theta} = 1$, hence $\mathbf{n}$ is a unit vector.
(b) $\frac{d}{dt}(\mathbf{n}) = (\cos\theta\mathbf{j} - \sin\theta\mathbf{i})\frac{d\theta}{dt}$
(c) $\frac{d}{dt}(\mathbf{n}).\mathbf{n} = (\cos\theta\mathbf{j} - \sin\theta\mathbf{i}).(\cos\theta\mathbf{i} + \sin\theta\mathbf{j})\frac{d\theta}{dt} = 0$.
2.15 Find the functions then differentiate.

(a)

$$\begin{aligned}\mathbf{r}.\mathbf{s} &= 2t^3 - 3t^2 - t - 2t^2 - t = 2t^3 - 5t^2 - 2t \\ \frac{d}{dt}(\mathbf{r}.\mathbf{s}) &= 6t^2 - 10t - 2\end{aligned}$$

(b)

$$\begin{aligned}\mathbf{r}\times\mathbf{s} &= \begin{vmatrix} \mathbf{i} & \mathbf{j} & \mathbf{k} \\ t^2 & -t & 2t+1 \\ 2t-3 & 1 & -t \end{vmatrix} \\ &= (t^2 - 2t - 1)\mathbf{i} + (t^3 + 4t^2 - 4t - 3)\mathbf{j} + (3t^2 - 3t)\mathbf{k} \\ \frac{d}{dt}(\mathbf{r}\times\mathbf{s}) &= (2t - 2)\mathbf{i} + (3t^2 + 8t - 4)\mathbf{j} + (6t - 3)\mathbf{k}\end{aligned}$$

(c)

$$\begin{aligned}|\mathbf{r}+\mathbf{s}| &= |(t^2+2t-3)\mathbf{i}+(1-t)\mathbf{j}+(t+1)\mathbf{k}| \\ &= [(t^2+2t-3)^2+(1-t)^2+(t+1)^2]^{1/2} \\ &= \left[t^4+4t^3-12t+11\right]^{1/2} \\ \frac{d}{dt}(|\mathbf{r}+\mathbf{s}|) &= \frac{2t^3+6t^2-6}{(t^4+4t^3-12t+11)^{1/2}}\end{aligned}$$

(d)

$$\mathbf{r}\times\frac{d}{dt}\mathbf{s} = \begin{vmatrix}\mathbf{i} & \mathbf{j} & \mathbf{k} \\ t^2 & -t & 2t+1 \\ 2 & 0 & -1\end{vmatrix} = t\mathbf{i}+(t^2+4t+2)\mathbf{j}+2t\mathbf{k}$$

$$\frac{d}{dt}(\mathbf{r}\times\frac{d}{dt}\mathbf{s}) = \mathbf{i}+(2t+4)\mathbf{j}+2\mathbf{k}$$

3.1 In each case multiply both numerator and denominator by the complex conjugate of the denominator.

$$\text{(a)}\ \frac{4-5\mathrm{i}}{1+2\mathrm{i}} = \frac{4-5\mathrm{i}}{1+2\mathrm{i}}\times\frac{1-2\mathrm{i}}{1-2\mathrm{i}} = \frac{-6-13\mathrm{i}}{1^2+2^2} = -\frac{6}{5}-\frac{13}{5}\mathrm{i}$$

$$\text{(b)}\ \frac{1+6\mathrm{i}}{7\mathrm{i}} = \frac{1+6\mathrm{i}}{7\mathrm{i}}\times\frac{-7\mathrm{i}}{-7\mathrm{i}} = \frac{42-7\mathrm{i}}{7^2} = \frac{42}{49}-\frac{1}{7}\mathrm{i}$$

3.2 Express the right hand side in the form $c+d\mathrm{i}$.

$$(2+5\mathrm{i})^2+\mathrm{i}(2-3\mathrm{i}) = 4-25+20\mathrm{i}+2\mathrm{i}+3 = -18+22\mathrm{i}$$

Equate real and imaginary parts: $a+b=-18 \quad a-b=22$.
These have the solution $a=2$, $b=-20$.

3.3 Multiply by complex conjugates to get real denominators.

$$\begin{aligned}z+\frac{1}{z} &= \frac{2+\mathrm{i}}{1-\mathrm{i}}+\frac{1-\mathrm{i}}{2+\mathrm{i}} = \frac{2+\mathrm{i}}{1-\mathrm{i}}\times\frac{1+\mathrm{i}}{1+\mathrm{i}}+\frac{1-\mathrm{i}}{2+\mathrm{i}}\times\frac{2-\mathrm{i}}{2-\mathrm{i}} \\ &= \frac{1+3\mathrm{i}}{2}+\frac{1-3\mathrm{i}}{5} = \frac{5+15\mathrm{i}}{10}+\frac{2-6\mathrm{i}}{10} = \frac{7}{10}+\frac{9}{10}\mathrm{i}\end{aligned}$$

3.4 If $z=a+b\mathrm{i}$ then $\bar{z}=a-b\mathrm{i}$, $z^2=(a^2-b^2)+2ab\mathrm{i}$ and $\bar{z}^2=(a^2-b^2)-2ab\mathrm{i}$. Substitute into the equation

$$z^2+\bar{z}^2+2z = (a^2-b^2)+2(a+b\mathrm{i}) = 4+4\mathrm{i}$$

Equate real and imaginary parts: $a^2-b^2+2a=4 \quad 2b\mathrm{i}=4\mathrm{i}$.
Therefore $b=2$ and $a^2+2a-8=0$ giving $a=-4$ or $a=2$. Hence the two complex numbers which satisfy the equation are $z=-4+2\mathrm{i}$, $z=2+2\mathrm{i}$.

3.5 The modulus is $r=\sqrt{1^2+(-\sqrt{3})^2}=2$. The tangent of the argument is the imaginary part divided by the real part, $\theta=\tan^{-1}\left(\frac{-\sqrt{3}}{1}\right)=-60^o$. The polar form is $2\angle-60^o$.

3.6 $(5\angle 30^o)(3\angle 15^o) = 15\angle 45^o = 15(\cos 45^o + \mathrm{i}\sin 45^o) = \frac{15}{\sqrt{2}} + \mathrm{i}\frac{15}{\sqrt{2}}$

3.7
$$\frac{5\angle 320^o}{2\angle 35^o} = \frac{5}{2}\angle 285^o = \frac{5}{2}(\cos 285^o + \mathrm{i}\sin 285^o) = 0.647 - 2.415\mathrm{i}$$

3.8
$$\frac{3\mathrm{e}^{\mathrm{i}4\pi/5}}{3\mathrm{e}^{\mathrm{i}\pi/2}} = \mathrm{e}^{\mathrm{i}4\pi/5-\mathrm{i}\pi/2} = \mathrm{e}^{\mathrm{i}3\pi/10} = 0.588 + 0.809\mathrm{i}$$

3.9 We will need the multiple angle formulae

$$\cos(A+B) = \cos A\cos B - \sin A\sin B,\ \sin(A+B) = \sin A\cos B + \cos A\sin B$$

Expanding the brackets and using the above identities gives

$$\begin{aligned} z_1z_2 &= r_1r_2(\cos\theta_1 + \mathrm{i}\sin\theta_1)(\cos\theta_2 + \mathrm{i}\sin\theta_2) \\ &= r_1r_2(\cos\theta_1\cos\theta_2 - \sin\theta_1\sin\theta_2) + r_1r_2\mathrm{i}(\cos\theta_1\sin\theta_2 + \sin\theta_1\cos\theta_2) \\ &= r_1r_2\cos(\theta_1+\theta_2) + r_1r_2\mathrm{i}\sin(\theta_1+\theta_2) = r_1r_2\angle(\theta_1+\theta_2) \end{aligned}$$

3.10 We have $\cos\theta = \frac{1}{2}\left(\mathrm{e}^{\mathrm{i}\theta} + \mathrm{e}^{-\mathrm{i}\theta}\right)$, therefore

$$\begin{aligned} \cos^4\theta &= \frac{1}{16}\left(\mathrm{e}^{\mathrm{i}\theta} + \mathrm{e}^{-\mathrm{i}\theta}\right)^4 = \frac{1}{16}\left(\mathrm{e}^{4\mathrm{i}\theta} + 4\mathrm{e}^{\mathrm{i}\theta} + 6\mathrm{e}^0 + 4\mathrm{e}^{-\mathrm{i}\theta} + \mathrm{e}^{-4\mathrm{i}\theta}\right) \\ &= \frac{3}{8} + \frac{1}{16}\left(\mathrm{e}^{4\mathrm{i}\theta} + \mathrm{e}^{-4\mathrm{i}\theta}\right) + \frac{1}{4}\left(\mathrm{e}^{\mathrm{i}\theta} + \mathrm{e}^{-\mathrm{i}\theta}\right) = \frac{3}{8} + \frac{1}{8}\cos(4\theta) + \frac{1}{2}\cos(2\theta) \end{aligned}$$

3.11 If we can express $-15 - 8\mathrm{i}$ in the form $r(\cos\theta + \mathrm{i}\sin\theta)$ then the square roots are $\sqrt{r}(\cos\frac{\theta}{2} + \mathrm{i}\sin\frac{\theta}{2})$. The modulus is $r = \sqrt{(-15)^2 + (-8)^2} = 17$. The complex number has negative real and imaginary parts, therefore θ is in the third quadrant and $\frac{\theta}{2}$ will be in the second quadrant. We have $\cos\theta = \frac{-15}{17}$, $\sin\theta = \frac{-8}{17}$ and the identity $2\cos^2\frac{\theta}{2} - 1 = \cos\theta$. Hence

$$\cos\frac{\theta}{2} = \pm\sqrt{\frac{\cos\theta + 1}{2}} = \pm\sqrt{\frac{1}{17}}$$

But in the second quadrant $\frac{\theta}{2}$ is positive and its cosine is negative, therefore $\cos\frac{\theta}{2} = -\frac{1}{\sqrt{17}}$ and $\sin\frac{\theta}{2} = \frac{4}{\sqrt{17}}$. Hence roots are $-1 + 4\mathrm{i}$ and $1 - 4\mathrm{i}$.

3.12 Modulus and argument: $r = \sqrt{(-1)^2 + 1^2} = \sqrt{2}$, $\quad \theta = \tan^{-1}(-1) = \frac{3\pi}{4}$.

$$-1 + \mathrm{i} = \sqrt{2}\left(\cos\left(\frac{3\pi}{4} + 2k\pi\right) + \mathrm{i}\sin\left(\frac{3\pi}{4} + 2k\pi\right)\right)$$

Thus

$$(-1+\mathrm{i})^{1/3} = 2^{1/6}\left(\cos\frac{3\pi/4 + 2k\pi}{3} + \mathrm{i}\sin\frac{3\pi/4 + 2k\pi}{3}\right)$$

and roots are (substitute $k = 0, 1, 2$),

$$2^{1/6}\left(\cos\frac{\pi}{4} + \mathrm{i}\sin\frac{\pi}{4}\right),\ 2^{1/6}\left(\cos\frac{11\pi}{12} + \mathrm{i}\sin\frac{11\pi}{12}\right),\ 2^{1/6}\left(\cos\frac{19\pi}{12} + \mathrm{i}\sin\frac{19\pi}{12}\right)$$

4.1(a) $v = z + z^{-1}$, $\frac{dv}{dz} = 1 - z^{-2} = 1 - \frac{1}{z^2}$
(b) $y = x^{0.8} + x^{-0.2}$, $\frac{dy}{dx} = 0.8x^{-0.2} - 0.2x^{-1.2} = \frac{0.2}{x^{0.2}}\left(4 - \frac{1}{x}\right)$.
(c) $z = 2y^{1/2} - 4y^{-1/2}$, $\frac{dz}{dy} = y^{-1/2} + 2y^{-3/2} = \frac{1}{\sqrt{y}}\left(1 + \frac{2}{y}\right)$.
4.2(a) $\frac{dy}{dx} = -15\sin(5x) - 6e^{2x}$ (b) $\frac{dy}{dx} = \frac{6}{x} + 4x^{-3} = \frac{6}{x} + \frac{4}{x^3}$
(c) $\frac{dP}{dt} = -24 + 12\cosh(4t)$ (d) $Q = \sin^2 x + \cos^2 x = 1, \forall x \in \Re$, $\frac{dQ}{dx} = 0$.
4.3 $s = 40t - 4t^2$, $v = \frac{ds}{dt} = 40 - 8t$ $v = 0$ when $40 - 8t = 0$, $t = 5$ seconds.
4.4 $P = P_0 e^{-kh}$, $\frac{dP}{dh} = -P_0 k e^{-kh}$. When $h = 3000$, $\frac{dP}{dh} = -8.677$.
4.5 $\theta = 2\sin(3t)$, $\frac{d\theta}{dt} = 6\cos(3t) = 0$ when $\cos(3t) = 0$, i.e. $t = \frac{\pi}{6}$ seconds.
4.6 Let $u = 3x - \pi$ so that $y = \sin u$. Then $\frac{du}{dx} = 3$, $\frac{dy}{du} = \cos u$.

$$\frac{dy}{dx} = \frac{dy}{du}\frac{du}{dx} = 3\cos u = 3\cos(3x - \pi)$$

4.7 Let $u = 5 - 7x$ so that $y = \ln u$. Then $\frac{du}{dx} = -7$, $\frac{dy}{du} = \frac{1}{u}$.

$$\frac{dy}{dx} = \frac{dy}{du}\frac{du}{dx} = \frac{-7}{u} = \frac{-7}{5 - 7x}$$

4.8 Let $u = 3x^2 + 1$ so that $y = e^u$. Then $\frac{du}{dx} = 6x$, $\frac{dy}{du} = e^u$.

$$\frac{dy}{dx} = \frac{dy}{du}\frac{du}{dx} = 6xe^u = 6xe^{3x^2+1}$$

4.9 Let $u = \cos t$ so that $P = u^7$. Then $\frac{du}{dt} = -\sin t$, $\frac{dP}{du} = 7u^6$.

$$\frac{dP}{dt} = \frac{dP}{du}\frac{du}{dt} = -7\sin t\cos^6 t = -7\sin t\cos^6 t$$

4.10 Let $u = \sqrt{\frac{k}{m}}t - \frac{\pi}{2}$ so that $y = y_0 \sin u$. Then $\frac{du}{dt} = \sqrt{\frac{k}{m}}$, $\frac{dy}{du} = y_0 \cos u$.

$$\frac{dy}{dt} = \frac{dy}{du}\frac{du}{dt} = \sqrt{\frac{k}{m}}y_0\cos u = \sqrt{\frac{k}{m}}y_0\cos\left(\sqrt{\frac{k}{m}}t - \frac{\pi}{2}\right)$$

When $t = 0$, $v = 0$ m/s and when $t = 3$, $v = 11.98$ m/s.
4.11 Let $u = x^3$ and $v = \cos(5x + 2)$. Then $\frac{du}{dx} = 3x^2$, $\frac{dv}{dx} = -5\sin(5x + 2)$.
Substitute into the product rule

$$\begin{aligned}\frac{dy}{dx} &= v\frac{du}{dx} + u\frac{dv}{dx} = \cos(5x+2)(3x^2) + x^3(-5\sin(5x+2)) \\ &= 3x^2\cos(5x+2) - 5x^3\sin(5x+2)\end{aligned}$$

4.12 Let $u = (x + 6)^3$ and $v = e^{5x+2}$. Then $\frac{du}{dx} = 3(x + 6)^2$, $\frac{dv}{dx} = 5e^{5x+2}$.
Substitute into the product rule

$$\begin{aligned}\frac{dy}{dx} &= v\frac{du}{dx}+u\frac{dv}{dx} = e^{5x+2}(3(x+6)^2)+(x+6)^3(5e^{5x+2})\\ &= (5x+33)(x+6)^2e^{5x+2}\end{aligned}$$

4.13 Let $u = e^x$ and $v = \sin(2x-1)$. Then $\frac{du}{dx}= e^x$, $\frac{dv}{dx}= 2\cos(2x-1)$.
Substitute into the product rule

$$\begin{aligned}\frac{dy}{dx} &= v\frac{du}{dx}+u\frac{dv}{dx} = \sin(2x-1)(e^x)+e^x(2\cos(2x-1))\\ &= e^x(\sin(2x-1)+2\cos(2x-1))\end{aligned}$$

4.14 Let $u = x^2$ and $v = \ln x$. Then $\frac{du}{dx}= 2x$, $\frac{dv}{dx} = \frac{1}{x}$.
Substitute into the product rule

$$\frac{dy}{dx} = v\frac{du}{dx}+u\frac{dv}{dx} = 2x\ln x + x^2\frac{1}{x} = x(1+2\ln x)$$

4.15 Let $u = Cn$ and $v = e^{-nk}$. Then $\frac{du}{dn}= C$, $\frac{dv}{dn}= -ke^{-nk}$.
Substitute into the product rule

$$\frac{dy}{dn} = v\frac{du}{dn}+u\frac{dv}{dn} = (e^{-nk})(C)+(Cn)(-ke^{-nk}) = Ce^{-nk}(1-nk)$$

4.16 Let $u = e^{-3t}$ and $v = 0.4\cos(50t)+0.1\sin(50t)$.
Then $\frac{du}{dt}= -3e^{-3t}$, $\frac{dv}{dt}= 5\cos(50t)-20\sin(50t)$.
Substitute into the product rule

$$\begin{aligned}\frac{dI}{dt} &= v\frac{du}{dt}+u\frac{dv}{dt}\\ &= (0.4\cos(50t)+0.1\sin(50t))(-3e^{-3t})+(e^{-3t})(5\cos(50t)-20\sin(50t))\\ &= e^{-3t}(3.8\cos(50t)-20.3\sin(50t))\end{aligned}$$

When $t = 0.2$, $\frac{dI}{dt} = 4.311$.

4.17 Let $u = \ln(5+x)$ and $v = 5+x$. Then $\frac{du}{dx} = \frac{1}{5+x}$, $\frac{dv}{dx}= 1$.
Substitute into the quotient rule

$$\frac{dy}{dx} = \frac{v\frac{du}{dx}-u\frac{dv}{dx}}{v^2} = \frac{(5+x)\left(\frac{1}{5+x}\right)-(\ln(5+x))(1)}{(5+x)^2} = \frac{1-\ln(5+x)}{(5+x)^2}$$

When $x = -4$, $\frac{dy}{dx} = 1$.

4.18 Let $u = e^{2t}$ and $v = 1+t^2$. Then $\frac{du}{dt}= 2e^{2t}$, $\frac{dv}{dt}= 2t$.
Substitute into the quotient rule

$$\frac{dP}{dt} = \frac{v\frac{du}{dt}-u\frac{dv}{dt}}{v^2} = \frac{(1+t^2)(2e^{2t})-(e^{2t})(2t)}{(1+t^2)^2} = \frac{2e^{2t}(t^2-t+1)}{(1+t^2)^2}$$

4.19 Let $u = 200x$, $v = 0.64x^2 + 200x + 350$. Then $\frac{du}{dx} = 200$, $\frac{dv}{dx} = 1.28x + 200$.
Substitute into the quotient rule

$$\begin{aligned} \frac{d\epsilon}{dx} &= \frac{v\frac{du}{dx} - u\frac{dv}{dx}}{v^2} = \frac{(0.64x^2 + 200x + 350)(200) - (200x)(1.28x + 200)}{(0.64x^2 + 200x + 350)^2} \\ &= \frac{70000 - 128x^2}{(0.64x^2 + 200x + 350)^2} \end{aligned}$$

$\frac{d\epsilon}{dx} = 0$ when $70000 - 128x^2 = 0$, i.e. $x = 23.385$.

4.20 Let $u = E^2R$ and $v = (R+r)^2$. Then $\frac{du}{dR} = E^2$, $\frac{dv}{dR} = 2(R+r)$.
Substitute into the quotient rule

$$\begin{aligned} \frac{dP}{dR} &= \frac{v\frac{du}{dR} - u\frac{dv}{dR}}{v^2} = \frac{(R+r)^2(E^2) - (E^2R)(2(R+r))}{(R+r)^4} \\ &= \frac{E^2(r^2 - R^2)}{(R+r)^4} = \frac{E^2(r-R)}{(R+r)^3} \end{aligned}$$

$\frac{d\epsilon}{dx} = 0$ when $70000 - 128x^2 = 0$, i.e. $x = 23.385$.

4.21 Substitute into the product rule

$$\begin{aligned} \frac{dy}{dx} &= u\frac{d}{dx}(v^{-1}) + (v^{-1})\frac{d}{dx}(u) = u(-v^{-2})\frac{dv}{dx} + v^{-1}\frac{du}{dx} \\ &= v^{-2}\left(v\frac{du}{dx} - u\frac{dv}{dx}\right) = \frac{v\frac{du}{dx} - u\frac{dv}{dx}}{v^2} \end{aligned}$$

4.22 We have

$y = \sin(2x) - \cos(3x)$ $\qquad \frac{dy}{dx} = 2\cos(2x) + 3\sin(3x)$

$\frac{d^2y}{dx^2} = -4\sin(2x) + 9\cos(3x)$ $\qquad \frac{d^3y}{dx^3} = -8\cos(2x) - 27\sin(3x)$

4.23 Let $u = e^t$ and $v = \sin t$. Then $\frac{du}{dt} = e^t$, $\frac{dv}{dt} = \cos t$.
Substitute into the product rule

$$\frac{dy}{dt} = v\frac{du}{dt} + u\frac{dv}{dt} = (\sin t)(e^t) + e^t(\cos t) = e^t(\sin t + \cos t)$$

Use the product rule again.
Let $u = e^t$ and $v = \sin t + \cos t$. Then $\frac{du}{dt} = e^t$, $\frac{dv}{dt} = \cos t - \sin t$.

$$\frac{d^2y}{dt^2} = v\frac{du}{dt} + u\frac{dv}{dt} = (\sin t + \cos t)(e^t) + e^t(\cos t - \sin t) = 2e^t\cos t$$

When $x = 0$, $\frac{d^2y}{dt^2} = 2$.

4.24 Differentiating implicitly

$$\frac{d}{dx}\left(x^3 - xy^2\right) = 3x^2 - \left(y^2 + 2xy\frac{dy}{dx}\right) = 0 \Rightarrow \frac{dy}{dx} = \frac{3x^2 - y^2}{2xy}$$

4.25 Differentiating implicitly

$$\frac{d}{dx}\left(\sin(xy) - x - 6\right) = \left(y + x\frac{dy}{dx}\right)\cos(xy) - 1 = 0 \Rightarrow \frac{dy}{dx} = \frac{1}{x\cos(xy)} - \frac{y}{x}$$

4.26 Taking logs of both sides:

$$\ln f = \frac{3}{2}\ln(x+7) + \ln(x-2) - 8\ln(2x-3) \Rightarrow \frac{1}{f}f' = \frac{\frac{3}{2}}{x+7} + \frac{1}{x-2} - \frac{16}{2x-3}$$

$$f'(x) = \frac{(x+7)^{3/2}(x-2)}{(2x-3)^8}\left(\frac{\frac{3}{2}}{x+7} + \frac{1}{x-2} - \frac{16}{2x-3}\right)$$

The only contribution to the value of the derivative at $x = 2$ is from the middle term, which cancels the $(x-2)$ multiplier in f. Hence $f'(2) = \frac{9^{3/2}}{1^8} = 27$.
4.27 Taking logs of both sides we have

$$\begin{aligned}\ln f &= \ln(2x+3)^x = x\ln(2x+3) \Rightarrow \frac{1}{f}f' = \ln(2x+3) + \frac{2x}{2x+3}\\ f'(x) &= (2x+3)^x\left(\ln(2x+3) + \frac{2x}{2x+3}\right)\end{aligned}$$

5.1 Straightforward integration of a power of x and a sine function.

$$\int 2x - \sin(3x)\, dx = \int 2x\, dx - \int \sin(3x)\, dx = x^2 + \frac{1}{3}\cos(3x) + c$$

5.2 Taking out the factor of 2 in the denominator leaves I9 in Table 5.1.

$$\int \frac{1}{2x^2+2}\, dx = \int \frac{1}{2(x^2+1)}\, dx = \frac{1}{2}\int \frac{1}{x^2+1}\, dx = \frac{1}{2}\tan^{-1} x + c$$

5.3 Integration of a power but be careful with the sign on the power.

$$\int \frac{4}{u^5}\, du = \int 4u^{-5}\, dt = -u^{-4} + c = c - \frac{1}{u^4}$$

5.4 Straightforward integration of an exponential.

$$\int_0^1 6e^{3t}\, dt = \left[2e^{3t}\right]_0^1 = 2e^3 - 2e^0 = 2(e^3 - 1)$$

5.5 The integration is

$$\int_0^x 2e^{2t}\, dt = \left[e^{2t}\right]_0^x = e^{2x} - e^0 = e^{2x} - 1$$

Equating this with unity:

$$e^{2x} - 1 = 1 \Rightarrow e^{2x} = 2 \Rightarrow 2x = \ln 2 \Rightarrow x = 0.693$$

5.6 Integrating each side we have

$$\int_0^1 \sqrt{\frac{3}{x}}\, dx = [2\sqrt{3x}]_0^1 - 2\sqrt{3} \text{ and } \int_0^b \sqrt{x}\, dx = \left[\frac{2}{3}x^{\frac{3}{2}}\right]_0^b = \frac{2}{3}b^{\frac{3}{2}}$$

Equating these expressions

$$\frac{2}{3}b^{\frac{3}{2}} = 2\sqrt{3} \Rightarrow b^{\frac{3}{2}} = 3\sqrt{3} = 3^{\frac{3}{2}} \Rightarrow b = 3$$

5.7 Let $u = x^2 - 1$ then $du = 2x\, dx$

$$\int \frac{x}{x^2-1}\, dx = \frac{1}{2}\int \frac{1}{u}\, du = \frac{1}{2}\ln|u| + c = \frac{1}{2}\ln|x^2-1| + c$$

5.8 Let $u = 3x^2 - 1$ then $du = 6x\, dx$

$$\int x(3x^2-1)^{\frac{1}{2}}\, dx = \frac{1}{6}\int u^{\frac{1}{2}}\, du = \frac{1}{9}u^{\frac{3}{2}} + c = \frac{1}{9}(3x^2-1)^{\frac{3}{2}} + c$$

5.9 Let $u = \cos x$ then $du = -\sin x\, dx$

$$\int \frac{\sin x}{\cos^2 x}\, dx = \int -\frac{1}{u^2}\, du = \frac{1}{u} + c = \sec x + c$$

5.10 Let $u = x^2 - 15$ then $du = 2x\, dx$. Limits become $u = 1$ and $u = 49$

$$\int_4^8 \frac{x}{\sqrt{x^2-15}}\, dx = \frac{1}{2}\int_1^{49} u^{-\frac{1}{2}}\, du = \left[u^{\frac{1}{2}}\right]_1^{49} = 7 - 1 = 6$$

5.11 Let $u = \cos x$ then $du = -\sin x\, dx$ Limits become $u = 1$ and $u = 0$.

$$\int_0^{\pi/2} \cos^2 x \sin x\, dx = \int_1^0 -u^2\, du = \left[-\frac{1}{3}u^3\right]_1^0 = 0 - (-\frac{1}{3}) = \frac{1}{3}$$

5.12 Let $u = e^x + 1$ then $du = e^x\, dx$. Limits become $u = 2$ and $u = e + 1$.

$$\int_0^1 \frac{e^x}{(e^x+1)^2}\, dx = \int_2^{e+1} \frac{1}{u^2}\, du = \left[-\frac{1}{u}\right]_2^{e+1} = -\frac{1}{e+1} - (-\frac{1}{2}) = 0.231$$

5.13 Let $u = \sin x$ then $du = \cos x\, dx$. Limits become $u = 0$, $u = 1$.

$$\int_0^{\pi/2} \frac{\cos x}{4+\sin^2 x}\, dx = \int_0^1 \frac{1}{4+u^2}\, du = \left[\frac{1}{2}\tan^{-1}\left(\frac{u}{2}\right)\right]_0^1 = 0.232$$

5.14 Let $x = \tan\theta$ then $dx = \sec^2\theta\, d\theta$. Limits become $\theta = 0$, $\theta = \frac{\pi}{4}$.

$$\int_0^1 \frac{1}{(1+x^2)^{3/2}}\,dx = \int_0^{\pi/4} \frac{1}{(1+\tan^2\theta)^{3/2}} \sec^2\theta\,d\theta$$
$$= \int_0^{\pi/4} \frac{\sec^2\theta}{(\sec^2\theta)^{3/2}}\,d\theta = \int_0^{\pi/4} \frac{1}{\sec\theta}\,d\theta = \int_0^{\pi/4} \cos\theta\,d\theta = [\sin\theta]_0^{\pi/4} = \frac{1}{\sqrt{2}}$$

5.15 Let $x = \sin\theta$, then $dx = \cos\theta\,d\theta$. Limits become $\theta = 0$ and $\theta = \frac{\pi}{2}$.

$$\int_0^1 x^2\sqrt{1-x^2}\,dx = \int_0^{\pi/2} \sin^2\theta(\cos\theta)\,\cos\theta\,d\theta = \int_0^{\pi/2} \sin^2\theta\cos^2\theta\,d\theta$$

$$= \int_0^{\pi/2} \frac{1}{4}\sin^2(2\theta)\,d\theta = \int_0^{\pi/2} \frac{1}{8} - \frac{1}{8}\cos(4\theta)\,d\theta = \left[\frac{1}{8}\theta - \frac{1}{32}\sin(4\theta)\right]_0^{\pi/2} = \frac{\pi}{16}$$

5.16 Both numerator and denominator are degree 1, so we seek A and B:

$$\frac{3x+2}{4x-5} = A + \frac{B}{4x-5} \Rightarrow 3x+2 = A(4x-5) + B = 4Ax + (B-5A)$$

Compare coefficients: $3 = 4A$ and $2 = B - 5A \Rightarrow A = \frac{3}{4},\ B = \frac{23}{4}$.

$$\text{Thus}: \quad \frac{3x+2}{4x-5} = \frac{3}{4} + \frac{23}{4(4x-5)}$$

5.17 The numerator has degree 2 and the denominator has degree 1, so:

$$\begin{aligned} \frac{2x^2+3x-5}{7x+11} &= Ax + B + \frac{C}{7x+11} \\ 2x^2+3x-5 &= (Ax+B)(7x+11) + C = 7Ax^2 + (11A+7B)x + 11B + C \end{aligned}$$

Comparing coefficients: $2 = 7A$, $3 = 11A + 7B$, $-5 = 11B + C \Rightarrow A = \frac{2}{7}$, $B = -\frac{1}{49}$, $C = -\frac{234}{49}$.

$$\frac{2x^2+3x-5}{7x+11} = \frac{2}{7}x - \frac{1}{49} - \frac{234}{49(7x+11)}$$

5.18 The numerator has degree 3 and the denominator has degree 1, so:

$$\begin{aligned} \frac{3x^3-8x^2+5x-1}{3x+4} &= Ax^2 + Bx + C + \frac{D}{3x+4} \\ 3x^3-8x^2+5x-1 &= 3Ax^3 + (4A+3B)x^2 + (4B+3C)x + 4C + D \end{aligned}$$

Comparing coefficients: $3 = 3A$, $-8 = 4A + 3B$, $5 = 4B + 3C$, $-1 = 4C + D$ $\Rightarrow A = 1,\ B = -4,\ C = 7,\ D = -29$.

$$\frac{3x^3-8x^2+5x-1}{3x+4} = x^2 - 4x + 7 - \frac{29}{3x+4}$$

5.19 The numerator has degree 4 and the denominator has degree 2, so:

$$\begin{aligned}\frac{3x^4}{x^2+2} &= Ax^2+Bx+C+\frac{Dx+E}{x^2+2}\\ 3x^4 &= Ax^4+Bx^3+(2A+C)x^2+(2B+D)x+(2C+E)\end{aligned}$$

Comparing coefficients: $3 = A$, $0 = B$, $0 = 2A + C$, $0 = 2B + D$, $0 = 2C + E$ $\Rightarrow A = 3$, $B = D = 0$, $C = -6$, $E = 12$.

$$\frac{3x^4}{x^2+2} = 3x^2 - 6 + \frac{12}{x^2+2}$$

5.20 Both numerator and denominator are degree 3, so:

$$\begin{aligned}\frac{7x^3}{(x-1)(x+2)(x-4)} &= A+\frac{Bx^2+Cx+D}{(x-1)(x+2)(x-4)}\\ 7x^3 &= A(x-1)(x+2)(x-4)+Bx^2+Cx+D\end{aligned}$$

Comparing the coefficients of x^3 gives $7 = A$. Substituting $x = 1$, $x = -2$, $x = 4$ gives: $7 = B + C + D$, $-56 = 4B - 2C + D$, $448 = 16B + 4C + D \Rightarrow B = 21$, $C = 42$, $D = -56$.

$$\frac{7x^3}{(x-1)(x+2)(x-4)} = 7 + \frac{21x^2+42x-56}{(x-1)(x+2)(x-4)}$$

5.21 The numerator has degree 3 and the denominator has degree 2, so:

$$\begin{aligned}\frac{4x^3+11x^2-9x-17}{(x-1)(x+2)} &= Ax+B+\frac{Cx+D}{(x-1)(x+2)}\\ 4x^3+11x^2-9x-17 &= Ax^3+(A+B)x^2+(B-2A+C)x+D-2B\end{aligned}$$

Comparing coefficients: $4 = A$, $11 = A+B$, $-9 = B-2A+C$, $-17 = -2B+D$ gives $A = 4$, $B = 7$, $C = -8$, $D = -3$.

$$\frac{4x^3+11x^2-9x-17}{(x-1)(x+2)} = 4x+7-\frac{8x+3}{(x-1)(x+2)}$$

5.22 We seek constants A and B such that

$$\begin{aligned}\frac{2x+3}{(x-1)(x-2)} &= \frac{A}{x-1}+\frac{B}{x-2}\\ 2x+3 &= A(x-2)+B(x-1)\end{aligned}$$

Substitute $x = 2 \Rightarrow 7 = 0 + B$ and substitute $x = 1 \Rightarrow 5 = -A$. So

$$\int \frac{2x+3}{(x-1)(x-2)}\,dx = \int \frac{-5}{x-1}+\frac{7}{x-2}\,dx = 7\ln|x-2| - 5\ln|x-1| + c$$

5.23 We seek constants A and B such that

$$\begin{aligned}\frac{7}{(x-1)(x+2)} &= \frac{A}{x-1}+\frac{B}{x+2}\\ 7 &= A(x+2)+B(x-1)\end{aligned}$$

Substitute $x=-2 \Rightarrow 7=0-3B$ and substitute $x=1 \Rightarrow 7=3A$. So

$$\int \frac{7}{(x-1)(x+2)}\,dx = \int \frac{7}{3(x-1)}-\frac{7}{3(x+2)}\,dx = \frac{7}{3}\ln\left|\frac{x-1}{x+2}\right|+c$$

5.24 We seek constants A and B such that

$$\begin{aligned}\frac{5x+4}{x(x+5)} &= \frac{A}{x}+\frac{B}{x+5}\\ 5x+4 &= A(x+5)+Bx\end{aligned}$$

Substitute $x=-5 \Rightarrow -21=0-5B$ and substitute $x=0 \Rightarrow 4=5A$. So

$$\int \frac{5x+4}{x(x+5)}\,dx = \int \frac{4}{5x}+\frac{21}{5(x+5)}\,dx = \frac{4}{5}\ln|x|+\frac{21}{5}\ln|x+5|+c$$

5.25 We seek constants A and B such that

$$\begin{aligned}\frac{-3}{x^2+2x-8}=\frac{-3}{(x+4)(x-2)} &= \frac{A}{x+4}+\frac{B}{x-2}\\ -3 &= A(x-2)+B(x+4)\end{aligned}$$

Substitute $x=2 \Rightarrow -3=0+6B$ and substitute $x=-4 \Rightarrow -3=-6A$. So

$$\int \frac{-3}{x^2+2x-8}\,dx = \int \frac{1}{2(x+4)}-\frac{1}{2(x-2)}\,dx = \frac{1}{2}\ln\left|\frac{x+4}{x-2}\right|+c$$

5.26 We seek constants A, B and C such that

$$\frac{-13x-19}{(x+1)(x-2)(x+3)} = \frac{A}{x+1}+\frac{B}{x-2}+\frac{C}{x+3}$$

$$-13x-19 = A(x-2)(x+3)+B(x+1)(x+3)+C(x+1)(x-2)$$

Substitute $x=2 \Rightarrow -45=15B$, substitute $x=-3 \Rightarrow 20=10C$ and substitute $x=-1 \Rightarrow -6=-6A$. So

$$\int \frac{-13x-19}{(x+1)(x-2)(x+3)}\,dx = \int \frac{1}{x+1}-\frac{3}{x-2}+\frac{2}{x+3}\,dx$$

$$= \ln|x+1|-3\ln|x-2|+2\ln|x+3|+c$$

5.27 We seek constants A, B and C such that

$$\begin{aligned}\frac{8-8x-x^2}{x(x-1)(x-2)} &= \frac{A}{x}+\frac{B}{x-1}+\frac{C}{x-2}\\ 8-8x-x^2 &= A(x-1)(x-2)+Bx(x-2)+Cx(x-1)\end{aligned}$$

Substituting $x = 0$, $x = 1$ and $x = 2$ in turn gives $8 = 2A$, $-1 = -B$ and $-12 = 2C$. So $A = 4$, $B = 1$ and $C = -6$,

$$\begin{aligned}\int \frac{8 - 8x - x^2}{x(x-1)(x-2)}\,dx &= \int \frac{4}{x} + \frac{1}{x-1} - \frac{6}{x-2}\,dx \\ &= 4\ln|x| + \ln|x-1| - 6\ln|x-2| + c\end{aligned}$$

5.28 We seek constants A, B, C and D such that

$$\begin{aligned}\frac{x^3 + x^2 + x + 2}{(x^2+1)(x^2+2)} &= \frac{Ax + B}{x^2+1} + \frac{Cx + D}{x^2+2} \\ x^3 + x^2 + x + 2 &= (Ax + B)(x^2+2) + (Cx + D)(x^2+1)\end{aligned}$$

Comparing coefficients of powers of x gives $1 = A + C$, $1 = B + D$, $1 = 2A + C$, $2 = 2B + D$. So $A = 0$, $B = 1$, $C = 1$ and $D = 0$.

$$\int \frac{x^3 + x^2 + x + 2}{(x^2+1)(x^2+2)}\,dx = \int \frac{1}{x^2+1} + \frac{x}{x^2+2}\,dx = \tan^{-1} x + \frac{1}{2}\ln(x^2+2) + c$$

5.29 We seek constants A, B and C such that

$$\begin{aligned}\frac{3x+5}{(x+1)(x-1)^2} &= \frac{A}{x+1} + \frac{B}{x-1} + \frac{C}{(x-1)^2} \\ 3x + 5 &= A(x-1)^2 + B(x+1)(x-1) + C(x+1)\end{aligned}$$

Substituting $x = 1$ and $x = -1$ gives the values of A and C directly, $A = \frac{1}{2}$ and $C = 4$. Comparing the coefficients of x^2 gives $0 = A + B$, hence $B = -\frac{1}{2}$. So

$$\begin{aligned}\int \frac{3x+5}{(x+1)(x-1)^2}\,dx &= \int \frac{1}{2(x+1)} - \frac{1}{2(x-1)} + \frac{4}{(x-1)^2}\,dx \\ &= \frac{1}{2}\ln\left|\frac{x+1}{x-1}\right| - \frac{4}{x-1} + c\end{aligned}$$

5.30 Using the same process as the previous example gives

$$\begin{aligned}\frac{4}{(x+1)^2(2x-1)} &= \frac{A}{x+1} + \frac{B}{(x+1)^2} + \frac{C}{2x-1} \\ 4 &= A(x+1)(2x-1) + B(2x-1) + C(x+1)^2\end{aligned}$$

Substitute $x = -1$ and $x = \frac{1}{2}$ to get $4 = -3B$ and $4 = \frac{9}{4}C$. Comparing the coefficients of x^2 gives $0 = 2A + C$, hence $A = -\frac{8}{9}$. So

$$\begin{aligned}\int \frac{4}{(x+1)^2(2x-1)}\,dx &= \int \frac{16}{9(2x-1)} - \frac{8}{9(x+1)} - \frac{4}{3(x+1)^2}\,dx \\ &= \frac{8}{9}\ln\left|\frac{2x-1}{x+1}\right| + \frac{4}{3(x+1)} + c\end{aligned}$$

5.31 We are looking for the partial fractions

$$\begin{aligned}\frac{x+1}{x^2(x-1)} &= \frac{A}{x}+\frac{B}{x^2}+\frac{C}{x-1}\\ x+1 &= Ax(x-1)+B(x-1)+Cx^2\end{aligned}$$

Substitute $x = 0$ and $x = 1$ to get $1 = -B$ and $2 = C$. Comparing the coefficients of x^2 gives $0 = A + C$, hence $A = -2$ and so

$$\int \frac{x+1}{x^2(x-1)}\,dx = \int \frac{2}{x-1} - \frac{2}{x} + \frac{1}{x^2}\,dx = \ln\left(\frac{x-1}{x}\right)^2 - \frac{1}{x} + c$$

5.32 We seek constants A, B, C and D such that

$$\begin{aligned}\frac{2x^3+5x^2-x+3}{(x-1)^2(x+2)^2} &= \frac{A}{x-1}+\frac{B}{(x-1)^2}+\frac{C}{x+2}+\frac{D}{(x+2)^2}\\ 2x^3+5x^2-x+3 &= A(x-1)(x+2)^2+B(x+2)^2\\ &\quad +C(x+2)(x-1)^2+D(x-1)^2\end{aligned}$$

Substituting $x = 1$ and $x = -2$ gives $9 = 9B$ and $9 = 9D$, i.e. $B = D = 1$. Comparing the coefficients of x^2 and x gives $A = C = 1$. Hence

$$\begin{aligned}\int \frac{2x^3+5x^2-x+3}{(x-1)^2(x+2)^2}\,dx &= \int \frac{1}{x-1}+\frac{1}{(x-1)^2}+\frac{1}{x+2}+\frac{1}{(x+2)^2}\,dx\\ &= \ln(x-1)(x+2) - \frac{1}{x-1} - \frac{1}{x+2} + c\end{aligned}$$

5.33 Since we cannot integrate $\tan^{-1} x$ directly we choose:

$$u = \tan^{-1} x,\ v' = 1 \quad \text{then} \quad u' = \frac{1}{x^2+1},\ v = x$$

$$\int \tan^{-1} x\,dx = x\tan^{-1} x - \int \frac{x}{x^2+1}dx = x\tan^{-1} x - \frac{1}{2}\ln(x^2+1) + c$$

5.34 We cannot integrate $\ln(x^2+4)$ directly so we choose

$$u = \ln(x^2+4),\ v' = 1 \quad \text{then} \quad u' = \frac{2x}{x^2+4},\ v = x$$

$$\int \ln(x^2+4)\,dx = x\ln(x^2+4) - \int \frac{2x^2}{x^2+4}\,dx$$

$$= \quad x\ln(x^2+4) - \int 2 - \frac{8}{x^2+4}dx = x\ln(x^2+4) - 2x + 4\tan^{-1}\left(\frac{x}{2}\right) + c$$

5.35 If we apply the technique of integration by parts with $u = 3t^2$ we will get an integral on the right hand side which will be a multiple of te^t. Applying the technique a second time will reduce it to a standard integral. Let

$$u = 3t^2,\ v' = e^t \quad \text{then} \quad u' = 6t,\ v = e^t$$

$$\int uv'dt = uv - \int vu'dt = 3t^2e^t - \int 6te^t dt$$

Using integration by parts again we have

$$u = 6t,\ v' = e^t \quad \text{then} \quad u' = 6,\ v = e^t$$

$$\begin{aligned} \int 6te^t\, dt &= 6te^t - \int 6e^t dt = 6te^t - 6e^t + c \\ \text{Hence} \quad \int 3t^2e^t\, dt &= 3t^2e^t - 6te^t + 6e^t + c \end{aligned}$$

5.36 Let $u = e^{-6t}$, $v' = 10\sin(2t)$ then $u' = -6e^{-6t}$, $v = -5\cos(2t)$.

$$\int uv'dt = uv - \int vu'dt = -5e^{-6t}\cos(2t) - \int 30e^{-6t}\cos(2t)\, dt$$

Integrate by parts again. Let $u = e^{-6t}$, $v' = 30\cos(2t)$ then $u' = -6e^{-6t}$, $v = 15\sin(2t)$.

$$\int 10e^{-6t}\sin(2t)\, dt = -5e^{-6t}\cos(2t) - 15e^{-6t}\sin(2t) - \int 90e^{-6t}\sin(2t)\, dt$$

The integral on the right hand side is a multiple of the one on the left, therefore we take it over to the left hand side, giving

$$\begin{aligned} \int 100e^{-6t}\sin(2t)\, dt &= -5e^{-6t}\cos(2t) - 15e^{-6t}\sin(2t) \\ \int 10e^{-6t}\sin(2t)\, dt &= -\frac{1}{2}e^{-6t}(\cos(2t) + 3\sin(2t)) \end{aligned}$$

5.37 We choose $u = x$, $v' = e^x$ then $u' = 1$, $v = e^x$. Then

$$\int_0^1 xe^x\, dx = [xe^x]_0^1 - \int_0^1 e^x\, dx = [xe^x - e^x]_0^1 = (e - e) - (0 - 1) = 1$$

5.38 As we cannot integrate $\ln x$ directly we choose

$$u = \ln x \quad v' = 1 \text{ then } u' = \frac{1}{x} \quad v = x$$

$$\int_1^4 \ln x\, dx = [x\ln x]_1^4 - \int_1^4 1\, dx = [x\ln x - x]_1^4 = 4\ln 4 - 3 = 2.545$$

5.39 The integration is straightforward

$$\int_3^\infty \frac{1}{x^2}\, dx = \left[-\frac{1}{x}\right]_3^\infty = 0 - \left(-\frac{1}{3}\right) = \frac{1}{3}$$

5.40 At the beginning of the chapter we said the definite integral of a function represented the area between the function and the x–axis, hence

$$\int_3^6 -x^2\,dx = \left[-\frac{1}{3}x^3\right]_3^6 = -72-(-9) = -63$$

We do not expect a negative result because area is a positive quantity. In this example the curve is entirely *below* the x–axis leading to the negative value of the *integral.* The area is 63 square units.
5.41 The areas under the two curves y_1 and y_2 are

$$\int_a^b y_1\,dx \quad \text{and} \quad \int_a^b y_2\,dx$$

respectively. The area *between* the two curves is given by the difference of these two integrals. In the previous example we found the area bounded by the functions $y=-x^2$ and $y=0$ (the x–axis).
The two functions in this exercise intersect at the points where

$$2-x^2 = x-4\,,\text{i.e.} \quad 0 = x^2+x-6 \Rightarrow x=-3 \text{ and } x=2$$

The area enclosed by the curves is

$$\int_{-3}^2 (2-x^2)-(x-4)\,dx = \left[6x-\frac{1}{2}x^2-\frac{1}{3}x^3\right]_{-3}^2 = 20\frac{5}{6}\ \text{units}^2$$

5.42 We have

$$V=\int_0^4 \pi y^2\,dx = \int_0^4 \pi e^{-2x}\,dx = \left[-\frac{\pi}{2}e^{-2x}\right]_0^4 = -\frac{\pi}{2}e^{-8}+\frac{\pi}{2} = 1.570$$

5.43 Let $u=x$, $\quad v'=\cos(nx)$ then $u'=1$, $\quad v=\frac{1}{n}\sin(nx)$. Then

$$\begin{aligned} a_n &= \frac{1}{\pi}\left[\frac{1}{n}x\sin(nx)\right]_{-\pi}^{\pi} - \frac{1}{\pi}\int_{-\pi}^{\pi}\frac{1}{n}\sin(nx)\,dx \\ &= \frac{1}{\pi}\left[\frac{1}{n}x\sin(nx)+\frac{1}{n^2}\cos(nx)\right]_{-\pi}^{\pi} = 0 \end{aligned}$$

For b_n we use $u=x$, $v'=\sin(nx)$ then $u'=1$, $v=-\frac{1}{n}\cos(nx)$.

$$\begin{aligned} b_n &= \frac{1}{\pi}\left[-\frac{1}{n}x\cos(nx)\right]_{-\pi}^{\pi} + \frac{1}{\pi}\int_{-\pi}^{\pi}\frac{1}{n}\cos(nx)\,dx \\ &= \frac{1}{\pi}\left[-\frac{1}{n}x\cos(nx)+\frac{1}{n^2}\sin(nx)\right]_{-\pi}^{\pi} = -\frac{2}{n}\cos(n\pi) \end{aligned}$$

5.44 Rewrite the integrand as follows

$$\int tr^t\,dt = \int te^{\ln r^t}\,dt = \int te^{t\ln r}\,dt$$

Using integration by parts

$$\begin{aligned}\int te^{t\ln r}\,dt &= \frac{t}{\ln r}e^{t\ln r} - \int \frac{1}{\ln r}e^{t\ln r}\,dt = \frac{t}{\ln r}e^{t\ln r} - \frac{1}{(\ln r)^2}e^{t\ln r} + c\\ &= \left(\frac{t}{\ln r} - \frac{1}{(\ln r)^2}\right)e^{t\ln r} + c = \left(\frac{t}{\ln r} - \frac{1}{(\ln r)^2}\right)r^t + c\end{aligned}$$

6.1 Values of end points are $y(1) = 0$, $y(e) = 1$. Differentiating the function and applying the MVT

$$\frac{dy}{dx} = \frac{1}{x} \Rightarrow \frac{1}{x_0} = \frac{y(e) - y(1)}{e - 1} \Rightarrow x_0 = \frac{e-1}{1-0} = e - 1$$

6.2 Obviously true for $x > 1$ since 1 is max value of sine. Consider any interval $[0, x]$ where $0 < x < 1$. On such an interval $\sin x$ is continuous and differentiable so the MVT states:
If $f(x) = \sin x$ there exists a value x_0 in the interval $(0, x)$ such that

$$\sin x = \sin 0 + (x - 0)f'(x_0) = x\cos x_0$$

But, $\cos x_0 < 1$ therefore $x\cos x_0 < x$, hence $\sin x = x\cos x_0 < x$.
6.3 We have $f(l) = \frac{2\pi}{\sqrt{g}}l^{1/2}$ therefore $f'(l) = \frac{\pi}{\sqrt{g}}l^{-1/2}$.
Applying the MVT: $f(l+\epsilon) = f(l) + \epsilon f'(l_0)$ and so $f(l+\epsilon) - f(l) = \epsilon\frac{\pi}{\sqrt{l_0 g}}$ where $l < l_0 < l + \epsilon$ if $\epsilon > 0$ and $l + \epsilon < l_0 < l$ if $\epsilon < 0$.
6.4 The polynomial is a cubic so we can deduce certain features. The curve will cross the x-axis at most 3 times and have at most 2 stationary points.
Axis-crossing (touching):
Given $y = 0$ when $x = 1$ it must be possible to write the cubic in the form

$$x^3 - 2x + 1 = (x-1)(ax^2 + bx + c)$$

and comparing coefficients we find $x^3 - 2x + 1 = (x-1)(x^2 + x - 1)$. The curve crosses the x-axis when $x = 1$, $x \approx 0.62$, $x \approx -1.62$. When $x = 0$, $y = 1$.
Relative extrema: Find derivatives,

$$\frac{dy}{dx} = 3x^2 - 2 \qquad \frac{d^2y}{dx^2} = 6x$$

For a critical point $\frac{dy}{dx} = 0$, i.e. $3x^2 - 2 = 0$. Roots are $x = \pm 0.816$ and $y = -0.089$, $y = 2.089$ respectively. Find their nature using second derivative,

$$x = 0.816,\quad \frac{d^2y}{dx^2} > 0,\ \text{minimum}; \qquad x = -0.816,\quad \frac{d^2y}{dx^2} < 0,\ \text{maximum}\,.$$

Points of inflection: $\frac{d^2y}{dx^2} = 0 \Rightarrow 6x = 0$ and so $x = 0$, $y = 1$.
Endpoints: When $x = -2$, $y = -3$ and when $x = 3$, $y = 22$.

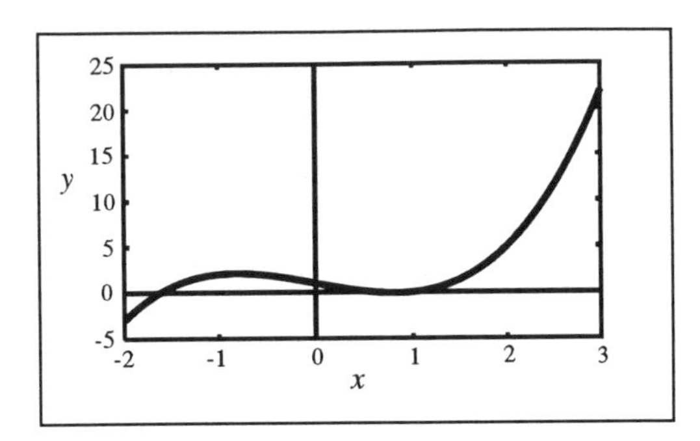

Fig. 10.2. Sketch for Exercise 6.4

6.5 As y can be written as a square we know y can never be negative and so will never cross the x-axis but may touch it.
Axis-crossing: When $x = 0$, $y = 16$ and $y = 0$ when $x^2 - 4 = 0 \Rightarrow x = \pm 2$. Curve crosses y-axis at (0,16) and touches x-axis at (-2,0) and (2,0).
Relative extrema: Find derivatives,

$$\frac{dy}{dx} = 4x^3 - 16x \qquad \frac{d^2y}{dx^2} = 12x^2 - 16 .$$

For a critical point $\frac{dy}{dx} = 0$, i.e. $4x^3 - 16x = 4x(x^2 - 4) = 0$. Roots are $x = 0$ and $x = \pm 2$. (Same points as axis crossing.)
Find their nature using the sign of the second derivative.

$$x = 0, \quad \frac{d^2y}{dx^2} = -16, \text{ maximum}; \qquad x = \pm 2, \quad \frac{d^2y}{dx^2} = 32, \text{ minimum} .$$

Points of inflection:

$$\frac{d^2y}{dx^2} = 12x^2 - 16 = 0 \Rightarrow x = \pm\frac{2}{\sqrt{3}}, \; y = \frac{64}{9} .$$

Endpoints: When $x = 3$, $y = 25$ and when $x = -3$, $y = 25$.

6.6 Let a and b be the numbers. The constraint is $a + b = 28$ and we want to maximise the function $y = a^3 + b^2$. We write $b = 28 - a$ and hence

$$y = a^3 + (28 - a)^2, \qquad \frac{dy}{da} = 3a^2 + 2a - 56, \qquad \frac{d^2y}{da^2} = 6a + 2 .$$

For a maximum/minimum $\frac{dy}{da} = 0$, i.e.

$$3a^2 + 2a - 56 = (3a + 14)(a - 4) = 0 \Rightarrow a = -14/3, \; a = 4 .$$

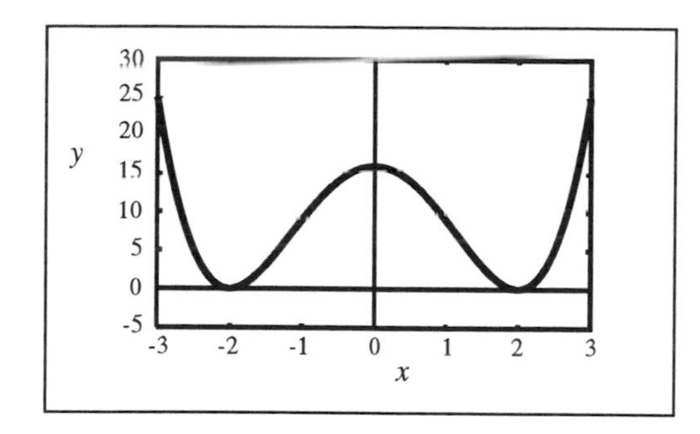

Fig. 10.3. Sketch for Exercise 6.5

Reject the negative value because the question specifies both a and b are non-negative. Thus $a = 4$ and $b = 28 - a = 24$. Check that we have a minimum using the sign of the second derivative: $\frac{d^2y}{da^2} = 6a + 2 = 26$, which is positive, hence we have found a minimum.

6.7 For a right angled triangle the longest side is the hypotenuse and the shorter sides represent the height and the base, hence our function is $A = \frac{1}{2}bh$ where b, h are the base and height satisfying $b + h = 20$ cm. Using $b = 20 - h$ gives

$$A = \frac{1}{2}bh = \frac{1}{2}h(20 - h) = 10h - \frac{1}{2}h^2, \qquad \frac{dA}{dh} = 10 - h, \qquad \frac{d^2A}{dh^2} = -1$$

The second derivative is always negative so any critical points will be maxima. For a maximum $\frac{dA}{dh} = 10 - h = 0$, hence $h = 10$ and $b = 20 - h = 10$. Maximum area is $A = \frac{1}{2}bh = 50$ cm^2.

6.8 The constraint is the volume which satisfies $V = \pi r^2 h = 500$ and the function to be minimised is the surface area $A = 2(2r)^2 + 2\pi r h$. Writing

$$\begin{aligned} h &= \frac{500}{\pi r^2} \Rightarrow A = 2(2r)^2 + 2\pi r h = 8r^2 + \frac{1000}{r} \\ \frac{dA}{dr} &= 16r - \frac{1000}{r^2}, \qquad \frac{d^2A}{dr^2} = 16 + \frac{2000}{r^3} \end{aligned}$$

For a critical point $\frac{dA}{dr} = 16r - \frac{1000}{r^2} = 0 \Rightarrow 16r^3 = 1000$, hence $r = 3.969$ cm. We then have $\frac{d^2A}{dr^2} = 16 + \frac{2000}{r^3} = 48$. This is positive, therefore $r = 3.969$ is a minimum and $A = 8r^2 + \frac{1000}{r} = 378$ cm^2 to the nearest square centimetre.

6.9 The perimeter is $P = 2h + 2r + \pi r = 2.5$ and the function to be minimised is the area $A = 2rh + \frac{1}{2}\pi r^2$. Writing $2h = 2.5 - 2r - \pi r$

$$\begin{aligned} A &= r(2.5 - 2r - \pi r) + \frac{1}{2}\pi r^2 = 2.5r - 2r^2 - \frac{\pi r^2}{2} \\ \frac{dA}{dr} &= 2.5 - 4r - \pi r \qquad \frac{d^2A}{dr^2} = -4 - \pi \end{aligned}$$

Second derivative is always negative therefore critical points are maxima. For a critical point $\frac{dA}{dr} = 0$.

$$0 = 2.5 - 4r - \pi r \Rightarrow r = \frac{2.5}{\pi + 4} = 0.350 \text{ m} .$$

Hence the maximum area is $A = 2.5r - 2r^2 - \frac{1}{2}\pi r^2 = 0.44 \text{ m}^2$.
6.10 The base of the box has area $A = (40 - 2x)(64 - 2x)$ and the height of the box will be x. Hence the volume is

$$V = x(40 - 2x)(64 - 2x) = 2560x - 208x^2 + 4x^3 .$$

(In this example the constraint is on the limiting values of x. The length and width of the box must be positive, i.e. $x < 20$ and $x < 32$.)

$$\frac{dV}{dx} = 2560 - 416x + 12x^2 \qquad \frac{d^2V}{dx^2} = 24x - 416 .$$

For a critical point $\frac{dV}{dx} = 0$.

$$x = \frac{416 \pm \sqrt{416^2 - 4 \times 12 \times 2560}}{24} .$$

Either $x = 8$ or $x = 26.667$. Reject the larger value because of constraint on x. Hence squares of 8 cm should be cut to give a maximum volume of

$$V = 8(40 - 16)(64 - 16) = 9216 \text{ cm}^3$$

6.11 (a) The radius is found using Pythagoras.

$$r^2 + h^2 = 6^2 \Rightarrow r^2 = 36 - h^2$$

(b) The volume of a cylinder with base radius r and height $2h$ is

$$\begin{aligned} V &= \pi r^2(2h) = 2\pi h(36 - h^2) = 72\pi h - 2\pi h^3 \\ \frac{dV}{dh} &= 72\pi - 6\pi h^2 \qquad \frac{d^2V}{dh^2} = -12\pi h \end{aligned}$$

(c) Second derivative is negative for positive h therefore critical points are maxima. For a critical point $\frac{dV}{dh} = 0$.

$$0 = 72\pi - 6\pi h^2 \Rightarrow h = \sqrt{12} .$$

h is assumed positive hence maximum volume of

$$V = 2\pi h(36 - h^2) = 48\pi\sqrt{12} = 522 \text{ cm}^3$$

6.12 Invert ϵ to get y.

$$y - \frac{0.64x^2 + 200x + 350}{200x} = \frac{0.64x^2}{200x} + \frac{200x}{200x} + \frac{350}{200x} = 1 + \frac{1.75}{x} + 0.0032x$$

Differentiating

$$\frac{dy}{dx} = 0.0032 - \frac{1.75}{x^2} \qquad \frac{d^2y}{dx^2} = \frac{3.5}{x^3}$$

Second derivative is positive for positive x therefore critical points are minima. For a critical point $\frac{dy}{dx} = 0$.

$$0 = 0.0032 - \frac{1.75}{x^2} \Rightarrow x = \sqrt{\frac{1.75}{0.0032}} = 23.385 \text{ amp}$$

Maximum efficiency is

$$y = 1 + \frac{1.75}{x} + 0.0032x = 1.1497\,, \qquad \epsilon = \frac{1}{y} = 0.870$$

6.13 Evaluate the derivatives at $x = 0$,

$$\begin{aligned} f(0) &= \sin 0 = 0 & f'''(0) = -\cos 0 = -1 \\ f'0) &= \cos 0 = 1 & f^{(4)}(0) = \sin 0 = 0 \\ f''(0) &= -\sin 0 = 0 \end{aligned}$$

and so

$$f(x) = \sin x = f(0) + xf'(0) + \frac{x^2}{2!}f''(0) + \ldots + \frac{x^n}{n!}f^{(n)}(0) + \ldots = \frac{x}{1!} - \frac{x^3}{3!} + \frac{x^5}{5!} - \ldots$$

6.14

$$f''(x) = -\frac{1}{(1+x)^2},\ f'(x) = \frac{1}{1+x},\ f'''(x) = \frac{2}{(1+x)^3},\ \ldots$$

Therefore

$$f(x) = \ln(1+x) = f(0) + xf'(0) + \frac{x^2}{2!}f''(0) + \ldots = \frac{x}{1} - \frac{x^2}{2} + \frac{x^3}{3} - \frac{x^4}{4} + \ldots$$

6.15

$$f(x) = e^{x/3}, \quad f'(x) = \frac{1}{3}e^{x/3}, \quad f''(x) = \frac{1}{3^2}e^{x/3}, \ \ldots$$

and so

$$f(3) = e,\ f'(3) = \frac{1}{3}e,\ f''(3) = \frac{1}{3^2}e,\ f'''(3) = \frac{1}{3^3}e,\ \ldots$$

Therefore

$$\begin{aligned} f(x) &= e^{x/3} = f(3) + (x-3)f'(3) + \ldots + \frac{(x-3)^n}{n!}f^{(n)}(3) + \ldots \\ &= e\left[1 + \frac{(x-3)}{3} + \frac{(x-3)^2}{3^2.2!} + \ldots\right] \end{aligned}$$

7.1 First treat y as a constant

$$\text{(a) } \frac{\partial u}{\partial x} = 2x, \quad \text{(b) } \frac{\partial u}{\partial x} = 6x^2 - 6xy + 4y^2, \quad \text{(c) } \frac{\partial u}{\partial x} = \frac{3}{3x+y}$$

then treat x as a constant

$$\text{(a) } \frac{\partial u}{\partial y} = 2y, \quad \text{(b) } \frac{\partial u}{\partial y} = -3x^2 + 8xy - 3y^2, \quad \text{(c) } \frac{\partial u}{\partial y} = \frac{1}{3x+y}$$

7.2 Find the required partial derivatives,

$$\begin{aligned} \frac{\partial z}{\partial x} &= 2e^{2x}\sin(3y) & \frac{\partial^2 z}{\partial x^2} &= 4e^{2x}\sin(3y) \\ \frac{\partial z}{\partial y} &= 3e^{2x}\cos(3y) & \frac{\partial^2 z}{\partial y^2} &= -9e^{2x}\sin(3y) \\ \frac{\partial^2 z}{\partial x \partial y} &= \frac{\partial^2 z}{\partial y \partial x} = 6e^{2x}\cos(3y) \end{aligned}$$

7.3

$$\begin{aligned} \frac{\partial u}{\partial x} &= 2x - 2y & \frac{\partial u}{\partial y} &= -2x + 2y \\ \frac{\partial u}{\partial x} + \frac{\partial u}{\partial y} &= (2x - 2y) + (-2x + 2y) = 0 \end{aligned}$$

7.4

$$\begin{aligned} f_x &= -e^{-x}\cos y - e^{-y}\sin x & f_{xx} &= e^{-x}\cos y - e^{-y}\cos x \\ f_y &= -e^{-x}\sin y - e^{-y}\cos x & f_{yy} &= -e^{-x}\cos y + e^{-y}\cos x \\ \text{Hence} \quad f_{xx} + f_{yy} &= \left(e^{-x}\cos y - e^{-y}\cos x\right) + \left(-e^{-x}\cos y + e^{-y}\cos x\right) = 0 \end{aligned}$$

Therefore f is harmonic.
7.5 Treat h as constant: $\frac{\partial V}{\partial r} = \frac{2}{3}\pi r h$. Treat r as constant: $\frac{\partial V}{\partial h} = \frac{1}{3}\pi r^2$.
7.6 Treat R as constant: $\frac{\partial P}{\partial E} = \frac{2E}{R}$. Treat E as constant: $\frac{\partial P}{\partial R} = -\frac{E^2}{R^2}$.
When $E = 200$ V and $R = 4000\Omega$

$$\frac{\partial P}{\partial E} = \frac{2E}{R} = \frac{1}{10} \text{ watts/volt} \qquad \frac{\partial P}{\partial R} = -\frac{E^2}{R^2} = \frac{1}{400} \text{ watts/ohm}$$

7.7 Treat C as constant: $\frac{\partial f}{\partial L} = -\frac{1}{4\pi\sqrt{L^3 C}}$.
Treat L as constant: $\frac{\partial f}{\partial C} = -\frac{1}{4\pi\sqrt{LC^3}}$.
When L=0.25 henrys and $C = 1.4 \times 10^{-4}$ farads

$$\frac{\partial f}{\partial L} = -\frac{1}{4\pi\sqrt{L^3 C}} = -53.8042 \qquad \frac{\partial f}{\partial C} = -\frac{1}{4\pi\sqrt{LC^3}} = 96078.9$$

7.8 First find the partial derivatives.

$$\frac{\partial z}{\partial x} = 4x - 3y \qquad \frac{\partial z}{\partial y} = -3x + 10y$$

Substitute the values into the total differential

$$\begin{aligned} dz &= \frac{\partial z}{\partial x}dx + \frac{\partial z}{\partial y}dy = (4x - 3y)dx + (10y - 3x)dy \\ &= (-2)(0.01) + (17)(-0.02) = -0.36 \end{aligned}$$

Calculate exact change: $z(1,2) - z(1.01, 1.98) = 12 - 11.6428 = -0.3572$.

7.9 First find the partial derivatives.

$$\frac{\partial n}{\partial l} = -\frac{1}{2rl^2}\sqrt{\frac{T}{\pi d}} \qquad \frac{\partial n}{\partial T} = \frac{1}{4rl\sqrt{\pi d T}}$$

Substitute the values into the total differential

$$dn = \frac{\partial n}{\partial l}dl + \frac{\partial n}{\partial T}dT = \frac{1}{4rl\sqrt{\pi d T}}dT - \frac{1}{2rl^2}\sqrt{\frac{T}{\pi d}}dl$$

7.10 Find all four partial derivatives.

$$\frac{\partial z}{\partial x} = 4x + 3y, \quad \frac{\partial z}{\partial y} = 3x - 8y, \quad \frac{dx}{dt} = \cos t, \quad \frac{dy}{dt} = -\sin t$$

Substitute the values into the total derivative

$$\frac{dz}{dt} = \frac{\partial z}{\partial x}\frac{dx}{dt} + \frac{\partial z}{\partial y}\frac{dy}{dt} = (4x + 3y)\cos t - (3x - 8y)\sin t$$

7.11 First find the partial derivatives of ϵ.

$$\frac{\partial \epsilon}{\partial R} = \frac{\alpha}{2}\sqrt{\frac{B}{R}} \qquad \frac{\partial \epsilon}{\partial B} = \frac{\alpha}{2}\sqrt{\frac{R}{B}}$$

Substitute the values into the total differential

$$\begin{aligned} \frac{d\epsilon}{dt} &= \frac{\partial \epsilon}{\partial R}\frac{dR}{dt} + \frac{\partial \epsilon}{\partial B}\frac{dB}{dt} \\ &= \frac{\alpha}{2}\sqrt{\frac{B}{R}}\frac{dR}{dt} + \frac{\alpha}{2}\sqrt{\frac{R}{B}}\frac{dB}{dt} = \frac{\alpha}{2\sqrt{BR}}\left(R\frac{dB}{dt} + B\frac{dR}{dt}\right) \end{aligned}$$

7.12 First find the partial derivatives.

$$\begin{aligned} f_x &= f_y = \frac{1}{1 + x + y} \qquad f_{xx} = f_{xy} = f_{yy} = \frac{-1}{(1 + x + y)^2} \\ f_{xxx} &= f_{xxy} = f_{xyy} = f_{yyy} = \frac{2}{(1 + x + y)^3} \end{aligned}$$

The expansion about $(0,0)$ is then

$$f(x,y) = f(0,0) + xf_x(0,0) + yf_y(0,0) + \frac{x^2 f_{xx} + 2xyf_{xy} + y^2 f_{yy}}{2!} + \ldots$$

At $x = y = 0$ we have

$$\ln(1+x+y) = \frac{x+y}{1!} - \frac{(x+y)^2}{2!} + \frac{(x+y)^3}{3!} \ldots$$

7.13 First find the partial derivatives.

$f_x = 2xe^{x^2+y}$ $\quad f_y = e^{x^2+y}$

$f_{xx} = (2+4x^2)e^{x^2+y}$ $\quad f_{yy} = e^{x^2+y}$ $\quad f_{xy} = 2xe^{x^2+y}$

$f_{xxx} = (12x+8x^3)e^{x^2+y}$ $\quad f_{xxy} = (2+4x^2)e^{x^2+y}$

$f_{xyy} = 2xe^{x^2+y}$ $\quad f_{yyy} = e^{x^2+y}$

At $x = y = 0$ we have

$$f(x,y) = 1 + \frac{y}{1!} + \frac{2x^2+y^2}{2!} + \frac{2x^2y+y^3}{3!} \ldots$$

7.14 Find first and second derivatives

$$f_x = 2x - 8,\ f_y = 2y,\ f_{xx} = 2,\ f_{xy} = 0,\ f_{xx} = 2$$

The first derivatives must be zero for critical points:

$$2x - 8 = 0 \quad \text{and} \quad 2y = 0$$

There is one critical point at $x = 4$, $y = 0$.
To determine its nature:

$$d = (f_{xy})^2 - f_{xx}f_{yy} = -4$$

d is negative and f_{xx} is positive, therefore $(4,0)$ is a relative minimum.
7.15 Find first and second derivatives

$$f_x = 4y - 4x^3,\ f_y = 4x - 4y^3,\ f_{xx} = -12x^2,\ f_{xy} = 4,\ f_{yy} = -12y^2$$

The first derivatives must be zero for critical points:

$$4y - 4x^3 = 0 \quad \text{and} \quad 4x - 4y^3 = 0$$

which gives

$$\begin{aligned} x - x^9 &= x(1-x^8) = 0 \\ y - y^9 &= y(1-y^8) = 0 \end{aligned}$$

These have the solutions $x = y = 0$, $x = y = \pm 1$.
To determine their nature create a table:

	(0,0)	(1,1)	(−1, −1)
f_{xx}	0	−12	−12
f_{xy}	4	4	4
f_{yy}	0	−12	−12
d	>0	<0	<0
	saddle	max	max

7.17 Eliminate x using constraint.

$$\begin{aligned} x &= 4 - 3y + 2z \\ f &= (4 - 3y + 2z)^2 + y^2 + z^2 = 16 + 10y^2 + 5z^2 - 12yz - 24y + 16z \end{aligned}$$

Find the partial derivatives.

$$\begin{aligned} f_y &= 20y - 12z - 24 \qquad\qquad f_z = 10z - 12y + 16 \\ f_{yy} &= 20 \qquad f_{yz} = -12 \qquad f_{zz} = 10 \end{aligned}$$

Can find nature of critical points now.

$$d = [f_{yz}]^2 - f_{yy}f_{zz} = -56$$

Negative d and positive f_{yy} indicates relative minima. First derivatives are zero for critical points.

$$\begin{aligned} 20y - 12z - 24 &= 0 \\ 10z - 12y + 16 &= 0 \end{aligned}$$

This pair of simultaneous equations has the solution $y = \frac{6}{7}$, $z = -\frac{4}{7}$ and $x = \frac{2}{7}$ is found from the constraint.

7.18 Let x be length, y the depth and z the height of the box. The circumference C and volume V are given by

$$V = xyz \qquad\qquad C = 2y + 2z$$

Therefore constraint is

$$x + 2y + 2z \leq 2$$

We seek a maximum so take equality and eliminate x from V.

$$\begin{aligned} x &= 2 - 2y - 2z \\ V &= (2 - 2y - 2z)yz = 2yz - 2y^2z - 2yz^2 \end{aligned}$$

Find the partial derivatives.

$$\begin{aligned} V_y &= 2z - 4yz - 2z^2 = 2z(1 - 2y - z) \\ V_z &= 2y - 2y^2 - 4yz = 2y(1 - y - 2z) \\ V_{yy} &= -4z \qquad V_{yz} = 2 - 4y - 4z \qquad V_{zz} = -4y \end{aligned}$$

First derivatives are zero for critical points.

$$2z(1-2y-z)=0 \quad \text{and} \quad 2y(1-y-2z)=0$$

This pair of simultaneous equations has the solution $y=\frac{1}{3}$, $z=\frac{1}{3}$ and $x=\frac{2}{3}$ is found from the constraint. Determine the nature.

$$d = [V_{yz}]^2 - V_{yy}V_{zz} = -\frac{4}{3}$$

Negative d and negative f_{yy} indicates relative maxima.

7.19 There is only one equality constraint so we are seeking one multiplier, λ_1. The problem now is to find the critical points of the function defined as

$$g = 4x_1x_2 + 6x_1x_3 + 4x_2x_3 + \lambda_1(x_1x_2x_3 - 12)$$

Find the four partial derivatives and set them to zero to find the critical points.

$$\begin{aligned}
\frac{\partial g}{\partial x_1} &= 4x_2 + 6x_3 + \lambda_1 x_2 x_3 = 0 \\
\frac{\partial g}{\partial x_2} &= 4x_1 + 4x_3 + \lambda_1 x_1 x_3 = 0 \\
\frac{\partial g}{\partial x_3} &= 6x_1 + 4x_2 + \lambda_1 x_1 x_2 = 0 \\
\frac{\partial g}{\partial \lambda_1} &= x_1x_2x_3 - 12 = 0
\end{aligned}$$

Eliminating λ_1 from first pair and the middle pair leads to.

$$\begin{aligned}
2x_3(3x_1 - 2x_2) &= 0 \\
2x_1(2x_2 - 3x_3) &= 0
\end{aligned}$$

Substituting into the constraint gives $x_1 = 2$, $x_2 = 3$ and $x_3 = 2$.

7.20 There is only one equality constraint so we are seeking one multiplier, λ_1. The problem now is to find the critical points of the function defined as

$$g = xy^2z^2 + \lambda_1(x+y+z-6)$$

Find the four partial derivatives and set them to zero to find the critical points.

$$\begin{aligned}
\frac{\partial g}{\partial x} &= y^2z^2 + \lambda_1 = 0 \\
\frac{\partial g}{\partial y} &= 2xyz^2 + \lambda_1 = 0 \\
\frac{\partial g}{\partial z} &= 2xy^2z + \lambda_1 = 0 \\
\frac{\partial g}{\partial \lambda_1} &= x + y + z - 6 = 0
\end{aligned}$$

Eliminating λ_1 from first pair and the middle pair leads to.

$$\begin{aligned} y^2z^2 - 2xyz^2 &= yz^2(y-2x) = 0 \\ 2xyz^2 - 2xy^2z &= 2xyz(z-y) = 0 \end{aligned}$$

Positivity requires $y = 2x$, $z = y$. So $x = \frac{6}{5}$, $y = \frac{12}{5}$ and $z = \frac{12}{5}$.

8.1

$$\begin{aligned} \int_0^1 \int_{x^2}^x xy\, dy\, dx &= \int_0^1 x\left[\frac{y^2}{2}\right]_{x^2}^x dx = \int_0^1 \frac{1}{2}x(x^2 - x^4)dx \\ &= \frac{1}{2}\int_0^1 x^3 - x^5 dx = \frac{1}{2}\left[\frac{x^4}{4} - \frac{x^6}{6}\right]_0^1 = \frac{1}{24} \end{aligned}$$

8.2

$$\begin{aligned} \int_0^1 \int_0^{2x^2} xe^y\, dy\, dx &= \int_0^1 x\left[e^y\right]_0^{2x^2} dx = \int_0^1 x(e^{2x^2} - 1)dx \\ &= \int_0^1 xe^{2x^2} - x dx = \left[\frac{1}{4}e^{2x^2} - \frac{1}{2}x^2\right]_0^1 = \frac{1}{4}e - \frac{3}{4} \end{aligned}$$

8.3 The intersection of the y functions is given by $x = 1$ and so the limits on x are $x = 0$ and $x = 1$. The partial derivatives of z are $\frac{\partial z}{\partial x} = 4$, $\frac{\partial z}{\partial y} = -3$. Then the surface area A is given by

$$\begin{aligned} A &= \int_0^1 \int_x^1 \sqrt{1 + 4^2 + (-3)^2}\, dy\, dx = \int_0^1 \int_x^1 \sqrt{26}\, dy\, dx \\ &= \sqrt{26}\int_0^1 [y]_x^1\, dx = \sqrt{26}\int_0^1 1 - x\, dx = \sqrt{26}\left[x - \frac{x^2}{2}\right]_0^1 = \frac{\sqrt{26}}{2} \end{aligned}$$

8.4 The integration is straightforward.

$$\begin{aligned} \int_0^1 \int_{x^2}^x \int_0^{xy} dz\, dy\, dx &= \int_0^1 \int_{x^2}^x [z]_0^{xy}\, dy\, dx = \int_0^1 \int_{x^2}^x xy\, dy\, dx \\ &= \int_0^1 \left[\frac{xy^2}{2}\right]_{x^2}^x dx = \int_0^1 \frac{1}{2}(x^3 - x^5)\, dx = \left[\frac{x^4}{8} - \frac{x^6}{12}\right]_0^1 = \frac{1}{24} \end{aligned}$$

8.5 Integrate with respect to z then y and finally x. The limits on z are 0 and $4 - x$. The limits on y are $\pm\sqrt{9 - x^2}$. The limits on x are ± 3.

$$\begin{aligned} \int_{-3}^3 \int_{-\sqrt{9-x^2}}^{\sqrt{9-x^2}} \int_0^{4-x} dz\, dy\, dx &= \int_{-3}^3 \int_{-\sqrt{9-x^2}}^{\sqrt{9-x^2}} [z]_0^{4-x}\, dy\, dx \\ = \int_{-3}^3 \int_{-\sqrt{9-x^2}}^{\sqrt{9-x^2}} 4 - x\, dy\, dx &= \int_{-3}^3 [(4-x)y]_{-\sqrt{9-x^2}}^{\sqrt{9-x^2}}\, dx \end{aligned}$$

$$= \int_{-3}^{3} (8-2x)\sqrt{9-x^2}\,dx = \int_{-3}^{3} 8\sqrt{9-x^2}\,dx - \int_{-3}^{3} 2x\sqrt{9-x^2}\,dx$$

$$= \int_{-3}^{3} 8\sqrt{9-x^2}\,dx + \left[\frac{2}{3}(9-x^2)^{3/2}\right]_{-3}^{3} = \int_{-3}^{3} 8\sqrt{9-x^2}\,dx + 0$$

$$= 8\int_{-\pi/2}^{\pi/2} 9\cos^2 u\,du = 36\int_{-\pi/2}^{\pi/2} 1+\cos(2u)\,du$$

$$= 36\left[u + \frac{1}{2}\sin(2u)\right]_{-\pi/2}^{\pi/2} = 36\pi$$

9.1 Order 1, degree 1, linear.

9.2 Order 3, degree 1, non-linear because $\cos(3y+\frac{\pi}{3})$ is not linear in y.

9.3 Order 2, degree 3, non-linear because e^{3P} is not a linear function of P.

9.4 Order 1, degree 1, linear.

9.5 $\frac{dy}{dx} = (2x+5)^3$, $y = \int (2x+5)^3\,dx = \frac{1}{8}(2x+5)^4 + C$.
$y(0) = 0$ so $0 = \frac{625}{8} + C$, and $y = \frac{1}{8}(2x+5)^4 - \frac{625}{8}$.

9.6 $\frac{dy}{dx} = \cos\left(3x+\frac{\pi}{3}\right)$, $y = \int \cos\left(3x+\frac{\pi}{3}\right)\,dx = \frac{1}{3}\sin\left(3x+\frac{\pi}{3}\right) + C$.
$y(0) = 0$ so $0 = \frac{1}{3}\frac{\sqrt{3}}{2} + C$, and $y = \frac{1}{3}\sin\left(3x+\frac{\pi}{3}\right) - \frac{1}{2\sqrt{3}}$.

9.7 $\frac{dy}{dx} = x^3+1$, $y = \int x^3+1\,dx = \frac{1}{4}x^4 + x + C$.
$y(2) = 6$ so $6 = 4+2+C$, and $y = \frac{1}{4}x^4 + x$.

9.8 $\frac{dy}{dx} = x^{\frac{3}{2}}$, $y = \int x^{\frac{3}{2}}\,dx = \frac{2}{5}x^{\frac{5}{2}} + C$.
$y(0) = 1$ so $1 = 0 + C$, and $y = \frac{2}{5}x^{\frac{5}{2}} + 1$.

9.9 $\frac{dy}{dx} = \cos x$, $y = \int \cos x\,dx = \sin x + C$.
$y(\frac{\pi}{2}) = 2$ so $2 = 1 + C$, and $y = \sin x + 1$.

9.10 $\frac{dy}{dx} = e^x$, $y = \int e^x\,dx = e^x + C$.
$y(0) = 1$ so $1 = 1 + C$, and $y = e^x$.

9.11 $\frac{di}{dt} = 3\sin\left(4t+\frac{\pi}{3}\right)$, $i = \int 3\sin\left(4t+\frac{\pi}{3}\right)\,dt = -\frac{3}{4}\cos\left(4t+\frac{\pi}{3}\right) + C$.
$i(0) = 0$ so $= \frac{3}{4}\cos\left(\frac{\pi}{3}\right)$, and $i = \frac{3}{8} - \frac{3}{4}\cos\left(4t+\frac{\pi}{3}\right)$.

9.12 The differential equation $\frac{dr}{dt} = 0.005$ has solution $r = 0.005t + C$. Substituting $r(0) = 2.1$ gives $C = 2.1$ and so the particular solution is $r = 2.1 + 0.005t$. The volume of a sphere is given by $V = \frac{4}{3}\pi r^3$ and the initial volume, V_0 is $V_0 = \frac{4}{3}\pi(2.1)^3$. The final volume is to be $1.1\times V_0$ so we solve for r

$$\frac{4}{3}\pi r^3 = 1.1 \times \frac{4}{3}\pi(2.1)^3 \;\Rightarrow\; r^3 = 1.1 \times (2.1)^3 \;\Rightarrow\; r = 2.1678 \text{ cm}.$$

We solve for t using: $2.1678 = 2.1 + 0.005t \;\Rightarrow\; t = 13.6$ sec .

9.13 General solution is $\theta = \int 1.5\cos\left(\frac{t}{6}\right)\,dt = 9\sin\left(\frac{t}{6}\right) + C$. The initial condition is $\theta(0) = 0$ so $C = 0$ and $\theta = 9\sin\left(\frac{t}{6}\right)$.

9.14 General solution is $P = \int -P_0 k e^{-kh}\,dh = P_0 e^{-kh} + C$.
Initial condition is $P(0) = P_0$, therefore $C = 0$ and so $P = P_0 e^{-kh}$. At $h = 3000$ metres $P = 1.01 \times 10^5 e^{-1.25\times10^{-4}\times3000} = 0.69 \times 10^5$ pascals.

9.15

$$\begin{aligned} E &= \int 1700\pi \cos(100\pi t) - 3600\pi \sin(100\pi t)\, dt \\ &= 17\sin(100\pi t) + 36\cos(100\pi t) + C \end{aligned}$$

Initial condition gives $5 = 0 + 36 + C \Rightarrow C = 9$, hence

$$E = 17\sin(100\pi t) + 36\cos(100\pi t) + 9 \text{ volts} .$$

9.16 $s = \int 40 - 5t\, dt = 40t - \frac{5}{2}t^2 + C$. Initial condition is $s(0) = 0$ therefore $C = 0$ and so $s = 40t - \frac{5}{2}t^2$. The car is stopped when $v = 0$, i.e. $0 = 40 - 5t \Rightarrow t = 8$ seconds. Thus $s = 160$ metres.

9.17 $k = 5$, $y = Ae^{5x}$.

9.18 $k = -2$, $y = Ae^{-2x}$.

9.19 $k = -\frac{1}{2}$, $y = Ae^{-\frac{1}{2}x}$.

9.20 $k = -\frac{1}{4}$, $x = Ae^{-\frac{1}{4}t}$.

9.21 (a) $k = -2$ $\quad y = Ae^{-2t}$, $4 = A$, $y = 4e^{-2t}$

(b) $k = -3$ $\quad x = Ae^{-3t}$, $2 = A$, $x = 2e^{-3t}$

(c) $k = -\frac{1}{6}$ $\quad y = Ae^{-\frac{1}{6}t}$, $7 = A$, $y = 7e^{-\frac{1}{6}t}$

(d) $k = -\frac{1}{4}$ $\quad i = Ae^{-\frac{1}{4}t}$, $1 = Ae^{-\frac{1}{2}}$, $i = e^{\frac{1}{2}-\frac{1}{4}t}$

9.22 The general solution is $F = Ae^{-kt}$. Initially none of the sugar is dissolved so $F(0) = 1$ which gives $A = 1$. After one minute the fraction undissolved is 0.75, i.e. $F(1) = 0.75$ and substituting this will give us the value of k.

$$0.75 = e^{-k} \Rightarrow k = -\ln 0.75 = 0.28768 \text{ (5 d.p.)}$$

To find the time it takes for half the sugar to dissolve we solve for t:

$$0.5 = e^{-kt} \Rightarrow -kt = \ln 0.5 \Rightarrow t = 2.41 \text{ min} .$$

9.23 The general solution to the differential equation is $N = Ae^{-\alpha t}$. We are not given an initial value of N so we denote it by N_0, i.e. $A = N_0$. What we are trying to do is find the time t for which $N = \frac{1}{2}N_0$, i.e. solve the following equation for t

$$N = N_0 e^{-\alpha t} = \frac{1}{2}N_0 \Rightarrow e^{-\alpha t} = 0.5 .$$

Taking logs gives $-\alpha t = \ln(0.5) \Rightarrow t = 0.509667 \times 10^{11}$ seconds

Convert this time from seconds to minutes to hours to days and then to years

$$t = 0.509667 \times 10^{11} \times \frac{1}{60} \times \frac{1}{60} \times \frac{1}{24} \times \frac{1}{365} = 1616 \text{ years.}$$

9.24 The general solution to the differential equation is $P = Ae^{kt}$ and the inital condition gives $A = 125000$. Our starting point was 1971 so 1981 corresponds to $t = 10$, therefore we solve for k:

$$139000 = 125000e^{10k} \Rightarrow k = \frac{1}{10}\ln\frac{139000}{125000} = 0.01062 \text{ (4 sig.fig.)}$$

Therefore the population in 2001 will be $P = 125000e^{0.01062\times 30} = 172000$.
9.25 The general solution is $P = Ae^{rt}$ and the initial condition fixes $A = P_0$ so the current deposit is $P = P_0e^{rt}$. The deposit will have trebled in value when $P = 3P_0$ and so we can solve for t.

$$3P_0 = P_0e^{rt} \Rightarrow e^{rt} = 3 \Rightarrow t = \frac{\ln 3}{0.075} = 14.65 \text{ years}.$$

9.26 The general solution is $P = Ae^{-kh}$ and the sea-level pressure gives $A = 1.01 \times 10^5$. To find the pressure at 5000 m we substitute $h = 5000$

$$P = 1.01 \times 10^5 e^{-1.25\times 10^{-4}\times 5000} = 0.54061 \times 10^5 \text{ pascals.}$$

The pressure at 5000 m is little more than half the pressure at sea-level, i.e. as you climb higher the air pressure drops.
9.27 Rearrange in the form $(y+1)^2dy = -x^3dx$ and integrating both sides gives the general solution $\frac{1}{3}(y+1)^3 = -\frac{1}{4}x^4 + C$, where C is an arbitrary constant. The neatest form of the solution is probably $4(y+1)^3 + 3x^4 = C$ but it is possible to get the explicit expression $y = \left(C - \frac{3}{4}x^4\right)^{1/3} - 1$.
9.28 Rewriting the differential equation in the form

$$\frac{dy}{y} = \frac{x^2}{1+x^3}\ dx$$

and then integrating, we get $\ln|y| = \frac{1}{3}\ln|1+x^3| + C$. Substituting the initial condition gives $C = \frac{2}{3}\ln 2$ and hence the general solution is $y^3 = 4(1+x^3)$.
9.29 Rearrange to get

$$\frac{dy}{\sqrt{1-y^2}} = 2x\ dx.$$

Integrating both sides gives the general solution $\sin^{-1} y = x^2 + C$. This can be rewritten in the more convenient form $y = \sin\left(x^2 + C\right)$. Substitute $y(0) = 0.5$ to get $0.5 = \sin C$. The principal value of C which satisfies this condition is $C = \frac{\pi}{6}$ but any value of the form $C = \frac{\pi}{6} + 2k\pi$ where k is an integer would suffice. The most convenient form of the particular solution is $y = \sin\left(x^2 + \frac{\pi}{6}\right)$.
9.30 Separating the variables gives

$$\int \frac{1}{E-v}\ dv = \int \frac{R}{L}\ dt \Rightarrow -\ln|E-v| = \frac{R}{L}t + C.$$

Substituting the initial condition $v(0) = 0$ gives $C = -\ln E$ and we have

$$\begin{aligned} \ln\left|\frac{E-v}{E}\right| &= -\frac{R}{L}t \Rightarrow \frac{E-v}{E} = e^{-\frac{R}{L}t} \\ v &= E\left(1 - e^{-\frac{R}{L}t}\right) = E\left(1 - e^{-\frac{t}{\tau}}\right) \end{aligned}$$

(a) After 0.15 seconds $v = 91(1 - e^{-0.30}) = 23.586$ volts.
(b) To find the value of t we go back to the solution involving logs

$$-\frac{R}{L}t = \ln\left|\frac{E-v}{E}\right| \Rightarrow t = -\tau \ln\left|\frac{E-v}{E}\right| = 0.450 \text{ seconds.}$$

9.31 The equation is separable and can be written in the form

$$\int \frac{dv}{g - kv} = \int dt \Rightarrow -\frac{1}{k}\ln|g - kv| = t + c\,.$$

Substituting the initial condition gives $c = -\frac{1}{k}\ln g$ and so

$$\begin{aligned} -\frac{1}{k}\ln|g - kv| &= t - \frac{1}{k}\ln g \Rightarrow 1 - \frac{kv}{g} = e^{-kt} \\ v &= \frac{g}{k}\left(1 - e^{-kt}\right) \end{aligned}$$

As $t \to \infty$ the exponential tends to zero and the limiting value of v is $\frac{g}{k}$.
9.32 The equation is separable and we have

$$\int \frac{dT}{T - T_a} = \int K\,dt$$

Integrating and substituting the initial condition $T(0) = 100\,^{o}$C gives

$$\ln|T - 30| = Kt + \ln|100 - 30| \Rightarrow \frac{T - 30}{70} = e^{Kt} \Rightarrow T = 30 + 70e^{Kt}$$

To determine K we use $T(15) = 70\,^{o}$C.

$$\ln|70 - 30| = 15K + \ln|100 - 30| \Rightarrow K = -0.0373 \text{ (4 d.p.)}$$

Finally to find t when $T = 40\,^{o}$C we have

$$\ln|40 - 30| = Kt + \ln|100 - 30| \Rightarrow t = \frac{1}{K}\ln\frac{10}{70} = 52.2 \text{ minutes.}$$

9.33 Rearranging this separable equation we have

$$\int \frac{di}{E - Ri} = \int \frac{dt}{L}$$

Integrating and substituting the initial condition $i(0) = 0$

$$\begin{aligned} -\frac{1}{R}\ln|E-Ri| &= \frac{t}{L}-\frac{1}{R}\ln E \\ \ln\left|\frac{E-Ri}{E}\right| &= -\frac{Rt}{L} \Rightarrow i=\frac{E}{R}\left(1-\mathrm{e}^{-\frac{Rt}{L}}\right) \end{aligned}$$

The limiting value of i is $\frac{E}{R}=\frac{1}{3}$ amps.
9.34 We have the variables separable problem

$$\frac{dV}{dt}=A-C_dV \qquad A=g\left(1-\frac{\rho_0}{\rho}\right)$$

Integrating gives

$$\int\frac{dV}{A-C_dV}=\int dt \Rightarrow -\frac{1}{C_d}\ln|A-C_dV|=t+C$$

Substituting the initial condition $V(0)=0$ gives $C=-\frac{1}{C_d}\ln|A|$, and hence

$$\ln\left|\frac{A-C_dV}{A}\right|=-C_dt \Rightarrow 1-\frac{C_d}{A}V=\mathrm{e}^{-C_dt} \Rightarrow V=\frac{A}{C_d}\left(1-\mathrm{e}^{-C_dt}\right)$$

(a) As the exponential term decays to zero, $V\to\frac{A}{C_d}=0.397$ m/s.
(b) To find t we revert to the stage in the solution involving logs

$$-C_dt=\ln\left|\frac{A-C_d}{A}\right| \Rightarrow t=-\frac{1}{C_d}\ln\left|1-\frac{C_dV}{A}\right|=0.167 \text{ seconds.}$$

9.35 Integrating factor: $p(x)=\mathrm{e}^{\int 1dx}=\mathrm{e}^x$. Equation becomes

$$\frac{d}{dx}\left(\mathrm{e}^xy\right)=\mathrm{e}^x\mathrm{e}^{-x}=1 \Rightarrow \mathrm{e}^xy=\int 1dx=x+C$$

Initial condition $y(0)=4$ gives $C=4$, hence $y=(x+4)\mathrm{e}^{-x}$.
9.36 Integrating factor: $p(x)=\mathrm{e}^{\int 1dx}=\mathrm{e}^x$. Equation becomes

$$\frac{d}{dx}\left(\mathrm{e}^xy\right)=(2+2x)\mathrm{e}^x \Rightarrow \mathrm{e}^xy=\int(2+2x)\mathrm{e}^x dx=2x\mathrm{e}^x+C$$

Initial condition $y(0)=2$ gives $C=2$, hence $y=2x+2\mathrm{e}^{-x}$.
9.37 Integrating factor:

$$p(t)=\mathrm{e}^{\int\frac{1}{t}+1dt}=\mathrm{e}^{t+\ln t}=t\mathrm{e}^t$$

Equation becomes

$$\frac{d}{dt}\left(t\mathrm{e}^tx\right)=3t^2\mathrm{e}^t \Rightarrow t\mathrm{e}^tx=\int 3t^2\mathrm{e}^tdt=3t^2\mathrm{e}^t-6t\mathrm{e}^t+6\mathrm{e}^t+C$$

Initial condition $x(0) = 6$ gives $C = -6$, hence $x = 3t - 6 + \frac{6}{t} - \frac{6}{t}e^{-t}$.

9.38 Integrating factor: $p(t) = e^{\int -6dt} = e^{-6t}$. Equation becomes

$$\frac{d}{dt}\left(e^{-6t}i\right) = 10e^{-6t}\sin(2t) \Rightarrow e^{-6t}i = \int 10e^{-6t}\sin(2t)dt$$

Integrate by parts (exercise 5.36) to get $e^{-6t}i = -\frac{1}{2}e^{-6t}(\cos(2t) - 3\sin(2t)) + C$. Initial condition $i(0) = 0$ gives $C = \frac{1}{2}$, hence $i = \frac{1}{2}e^{6t} - \frac{1}{2}\cos(2t) - \frac{3}{2}\sin(2t)$.

9.39 Auxiliary equation: $m^2 - 2m - 15 = (m-5)(m+3) = 0$.
Roots are $m = 5$ and $m = -3$. General solution is $y = Ae^{5x} + Be^{-3x}$.

9.40 Auxiliary equation: $m^2 - 4 = (m-2)(m+2) = 0$.
Roots are $m = 2$ and $m = -2$. General solution is $y = Ae^{2x} + Be^{-2x}$.

9.41 Auxiliary equation: $m^2 - 2m + 1 = (m-1)^2 = 0$. Repeated root is $m = 1$.
General solution is $y = (A + Bx)e^x$.

9.42 Auxiliary equation: $m^2 - 6m + 10 = 0$ has complex roots $m = 3 \pm i$.
General solution is $y = e^{3x}(C\cos x + D\sin x)$.

9.43 Auxiliary equation: $m^2 + 36 = 0$ has purely imaginary roots $m = \pm 6i$.
General solution is $y = C\cos(6x) + D\sin(6x)$.

9.44 Auxiliary equation: $m^2 - 3m + 2 = (m-2)(m-1) = 0$.
Roots $m = 2$ and $m = 1$. General solution is $y = Ae^{2x} + Be^x$.

9.45 Auxiliary equation: $m^2 + 100 = 0$ has roots $m = \pm 10i$.
General solution is $x = A\cos(10t) + B\sin(10t)$. Substituting the initial conditions $x(0) = 2$ and $\frac{dx}{dt}(0) = 0$ gives $2 = A$ and $0 = 10B$ so the solution is $x = 2\cos(10t)$.

9.46 Auxiliary equation: $m^2 + 7m + 12 = (m+4)(m+3) = 0$.
Roots $m = -4$ and $m = -3$. General solution is $x = Ae^{-4t} + Be^{-3t}$. Substituting the initial conditions $x(0) = 3$ and $\frac{dx}{dt}(0) = 6$ gives $3 = A + B$ and $6 = -4A - 3B$. The solution is $x = 18e^{-3t} - 15e^{-4t}$.

9.47 (a) Substitute the values of R, L and C to get

$$\frac{d^2i}{dt^2} + 500\frac{di}{dt} + 62500i = 0$$

Auxiliary equation: $m^2 + 500m + 62500 = (m + 250)^2 = 0$ has a repeated root $m = -250$. General solution is $i = (A + Bt)e^{-250t}$. Substituting the initial conditions $i(0) = 0$ and $\frac{di}{dt}(0) = 100$ give $0 = A$ and $100 = B - 250A$. The solution is $i = 100te^{-250t}$.

(b) Substitute the values of R, L and C to get

$$\frac{d^2i}{dt^2} + 1118\frac{di}{dt} + 62500i = 0$$

Auxiliary equation: $m^2 + 1118m + 62500 = 0$ has roots $m = -559 \pm 500$. General solution is $i = Ae^{-59t} + Be^{-1059t}$. Substituting $i(0) = 0$ and $\frac{di}{dt}(0) = 100$ gives

$0 = A + B$ and $100 = -59A - 1059B$, $A = 0.1$ and $B = -0.1$.
The solution is $i = 0.1\mathrm{e}^{-59t} - 0.1\mathrm{e}^{-1059t}$.
9.48 Auxiliary equation: $m^2 + 6m + 9 = (m + 3)^2 = 0$. Repeated root $m = -3$. General solution is $x = (A + Bt)\mathrm{e}^{-3t}$. Substituting $x(0) = 3$ and $\frac{dx}{dt}(0) = 6$ gives $3 = A$ and $6 = B - 3A$. The solution is $x = (3 + 15t)\mathrm{e}^{-3t}$.
9.49 Auxiliary equation: $m^2 + 2m + 10 = 0$ has roots $m = -1 \pm 3i$.
General solution is $\theta = \mathrm{e}^{-t}(C\cos(3t) + D\sin(3t))$. Substituting $\theta(0) = 0$ and $\frac{d\theta}{dt}(0) = 0$ gives $0 = C$ and $0.3 = 3D - C$. The solution is $\theta = 0.1\mathrm{e}^{-t}\sin(3t)$.
9.50 Auxiliary equation: $m^2 - 4m + 5 = 0$ has complex roots $m = 2 \pm i$.
Complementary function is $y_C = \mathrm{e}^{2x}(A\cos x + B\sin x)$.
Try the particular solution $y_P = C\mathrm{e}^x$ which gives $C = \frac{3}{2}$. General solution is $y = y_C + y_P = \mathrm{e}^{2x}(A\cos x + B\sin x) + \frac{3}{2}\mathrm{e}^x$.
9.51 Auxiliary equation: $m^2 + m - 6 = 0$ has roots $m = 2$ and $m = -3$.
Complementary function is $y_C = A\mathrm{e}^{2x} + B\mathrm{e}^{-3x}$.
Try the particular solution $y_P = Cx + D + E\mathrm{e}^{3x}$ then

$$9E\mathrm{e}^{3x}+3E\mathrm{e}^{3x}-6E\mathrm{e}^{3x}+C-6(Cx+D) = 3x+\mathrm{e}^{3x} \Rightarrow E = \frac{1}{6},\ C = -\frac{1}{2},\ D = -\frac{1}{12}$$

General solution is $y = y_C + y_P = A\mathrm{e}^{2x} + B\mathrm{e}^{-3x} - \frac{x}{2} - \frac{1}{12} + \frac{\mathrm{e}^{3x}}{6}$
9.52 Auxiliary equation: $m^2 - 2m = 0$ has roots $m = 2$ and $m = 0$.
Complementary function is $x_C = A\mathrm{e}^{2t} + B$.
Try the particular solution $x_P = C\cos(2t) + D\sin(2t)$ which gives

$$-4C\cos(2t) - 4D\sin(2t) + 4C\sin(2t) - 4D\cos(2t) = 3\cos(2t)$$

$$-4C - 4D = 3,\ 4C - 4D = 0$$

General solution is $x = x_C + x_P = A\mathrm{e}^{2t} + B - \frac{3}{8}(\cos(2t) + \sin(2t))$.
9.53 Auxiliary equation: $m^2 + 6m + 25 = 0$ has complex roots $m = -3 \pm 4i$.
Complementary function is $x_C = \mathrm{e}^{-3t}(A\cos(4t) + B\sin(4t))$.
Try particular solution $x_P = C\cos t + D\sin t$.

$$-C\cos t - D\sin t - 6C\sin t + 6D\cos t + 25C\cos t + 25D\sin t = 5\cos t$$

$$6D + 24C = 5,\quad -6C + 24D = 0 \Rightarrow C = \frac{20}{102},\ D = \frac{5}{102}$$

General solution is

$$x = x_C + x_P = \mathrm{e}^{-3t}(A\cos(4t) + B\sin(4t)) + \frac{20}{102}\cos t + \frac{5}{102}\sin t$$

9.54 Auxiliary equation: $m^2 + \frac{R}{L}m + \frac{1}{CL} = m^2 + 300m + 25000 = 0$.
Roots are $m = -150 \pm 50i$ and so $q_C = \mathrm{e}^{-150t}(A\cos(50t) + B\sin(50t))$. Substituting initial conditions gives $0 = A$ and $50 = -150A + 50B$. Therefore $q_C = \mathrm{e}^{-150t}\sin(50t)$. Try particular integral $q_P = C\cos(120t) + D\sin(120t)$.

$$
\begin{aligned}
- &\quad 14400C\cos(120t) - 14400D\sin(120t) - 36000C\sin(120t) \\
+ &\quad 36000D\cos(120t) + 25000C\cos(120t) + 25000D\sin(120t) = 11\cos(120t)
\end{aligned}
$$

$$10600D - 36000C = 0 \qquad 10600C + 36000D = 11$$

giving $C \approx 8.279 \times 10^{-5}$, $D \approx 2.812 \times 10^{-4}$. Thus general solution is

$$q = q_C + q_P = \mathrm{e}^{-150t}\sin(50t) + 8.279 \times 10^{-4}\cos 120t + 2.812 \times 10^{-3}\sin 120t$$

9.55 This is a first order differential equation which is variables separable

$$\int \frac{dp}{rp - W} = \int dt \Rightarrow \frac{1}{r}\ln|rp - W| = t + C$$

and substituting $p(0) = p_0$ we find C in terms of known constants.

$$
\begin{aligned}
\frac{1}{r}\ln|rp_0 - W| &= 0 + C \Rightarrow C = \frac{1}{r}\ln|rp_0 - W| \\
\text{and so} \quad \frac{1}{r}\ln\left|\frac{rp - W}{rp_0 - W}\right| &= t \Rightarrow rp - W = (rp_0 - W)\mathrm{e}^{rt} \\
p &= \left(p_0 - \frac{W}{r}\right)\mathrm{e}^{rt} + \frac{W}{r}
\end{aligned}
$$

The rest of the solution is exactly as presented in the introductory section. The initial capital, p_0, is £180,000. For this particular example the couple want $p(240) = 0$. So given the monthly interest rate, r, we calculate the monthly withdrawals to be

$$W = \frac{p_0 r \mathrm{e}^{240r}}{\mathrm{e}^{240r} - 1}$$

If the annual interest rate is equivalent to 0.5% per month the couple would be able to withdraw approximately £1288 per month to supplement their pension.
9.56 The equation looks complex but the solution is found by direct integration.

$$\frac{s}{k} - \frac{\pi}{4} = \int \frac{\pi R\theta}{2(h_0 + R\theta^2)}\, d\theta \pm \int \frac{R}{h_0 + R\theta^2}\, d\theta$$

The first term on the right hand side is integrated using a substitution. Let $u = 2h_0 + 2R\theta^2$ then $\frac{du}{d\theta} = 4R\theta$ and so $\frac{1}{4}du = R\theta d\theta$.

$$\int \frac{\pi R\theta}{2(h_0 + R\theta^2)}\, d\theta = \int \frac{\pi}{4u}\, du = \frac{\pi}{4}\ln u + c = \frac{\pi}{4}\ln(2h_0 + 2R\theta^2) + c$$

The second term on the right hand side is an inverse tan function.

$$\int \frac{R}{h_0 + R\theta^2}\, d\theta = \int \frac{1}{\frac{h_0}{R} + \theta^2}\, d\theta = \sqrt{\frac{R}{h_0}}\tan^{-1}\sqrt{\frac{R}{h_0}}\theta + c$$

The general solution(s) is

$$\frac{s}{k} - \frac{\pi}{4} = \frac{\pi}{4}\ln(h_0 + R\theta^2) \pm \sqrt{\frac{R}{h_0}}\tan^{-1}\sqrt{\frac{R}{h_0}}\theta + c\,.$$

To find the particular solutions we impose the initial condition and find the value of the constant of integration, c. When $s(\theta = 0) = \frac{\pi k}{4}$

$$\frac{\pi}{4} - \frac{\pi}{4} = \frac{\pi}{4}\ln h_0 + \sqrt{\frac{R}{h_0}}\tan^{-1}(0) + c \Rightarrow c = -\frac{\pi}{4}\ln h_0$$

and so the first solution is

$$\frac{s^+}{k} = \frac{\pi}{4} + \frac{\pi}{4}\ln\left(\frac{h_0 + R\theta^2}{h_0}\right) + \sqrt{\frac{R}{h_0}}\tan^{-1}\sqrt{\frac{R}{h_0}}\theta$$

When $s(\theta = \alpha) = \frac{\pi k}{4}$

$$\begin{aligned}\frac{\pi}{4} - \frac{\pi}{4} &= \frac{\pi}{4}\ln(h_0 + R\alpha^2) - \sqrt{\frac{R}{h_0}}\tan^{-1}\sqrt{\frac{R}{h_0}}\alpha + c \\ c &= \sqrt{\frac{R}{h_0}}\tan^{-1}\sqrt{\frac{R}{h_0}}\alpha - \frac{\pi}{4}\ln h_1\end{aligned}$$

and so the second solution is

$$\frac{s^-}{k} = \frac{\pi}{4} + \frac{\pi}{4}\ln\left(\frac{h_0 + R\theta^2}{h_1}\right) - \sqrt{\frac{R}{h_0}}\tan^{-1}\sqrt{\frac{R}{h_0}}\theta + \sqrt{\frac{R}{h_0}}\tan^{-1}\sqrt{\frac{R}{h_0}}\alpha$$

The neutral angle, θ_n, is found by equating these expressions for s^+ and s^-,

$$\frac{\pi}{4}\ln\left(\frac{h_1}{h_0}\right) + 2\sqrt{\frac{R}{h_0}}\tan^{-1}\sqrt{\frac{R}{h_0}}\theta_n - \sqrt{\frac{R}{h_0}}\tan^{-1}\sqrt{\frac{R}{h_0}}\alpha = 0$$

The explicit equation for θ_n is given by

$$\theta_n = \sqrt{\frac{h_0}{R}}\tan\left[\frac{\sqrt{\frac{R}{h_0}}\tan^{-1}\sqrt{\frac{R}{h_0}}\alpha - \frac{\pi}{4}\ln\left(\frac{h_1}{h_0}\right)}{2\sqrt{\frac{R}{h_0}}}\right]$$

Index

absolute extrema, 86
anti-derivative, 57
Argand diagram, 35
auxiliary equation, 171

basis, 24

characteristic equation, 171
complementary function, 170, 181
complex conjugate, 34
complex modulus, 34
constraint, 96
Cramer's rule, 3
critical point, 83
cross derivative, 108
cross product, 26
curve sketching, 93

d'Alembert's ratio test, 102
De Moivre, 40
dependent variable, 43
derivative, 44
determinant, 2
differentiable, 11
differential, 110
differential equation
– degree, 142
– general solution, 145
– homogenous, 170
– initial condition, 145
– linearity, 142
– order, 142
– particular solution, 145
– variables separable, 159
differentiation
– chain rule, 47
– function of a function, 47
– higher derivatives, 53
– implicit, 54
– logarithmic, 55
– product rule, 49
– quotient rule, 51
discriminant, 171
domain, 44
dot product, 25
double integral, 130

Euler's formula, 39
exponential form, 39

first derivative test, 83
Fourier coefficients, 79
function
– concavity, 84
– continuous, 44
– decreasing, 82
– increasing, 82
– stationary, 82

Gaussian elimination, 8
growth and decay, 154

harmonic, 109

ill–conditioning, 10
imaginary unit, 33
independent variable, 43
integrand, 57
integrating factor, 165
integration
– by parts, 77
– by substitution, 63
– constant of, 57
– definite, 58
– indefinite, 57
– limits, 59
– partial fractions, 71

Jacobian, 135

Lagrange multipliers, 124

linear independence, 24
linearly independent, 170

Maclaurin series, 102
matrix, 11
– adjoint, 18
– co–factors, 17
– column, 11
– determinant, 16
– identity, 16
– inverse, 18
– minors, 17
– multiplication, 13
– row, 11
– square, 11
– transpose, 12
Mean value theorem, 89

Newton's second law, 139

optimisation, 96

parallelepiped, volume, 28
partial derivative, 106
partial differential, 110
partial fraction
– improper, 68
– proper, 68
– repeated, 74
– resolving, 71
partial fractions, 67
particular integral, 170
point of inflection, 86
polar form, 37
principal argument, 37
principle of superposition, 171

radius of convergence, 103
range, 44
relative extrema, 83, 117
relative maximum, 83
relative minimum, 83
remainder, Cauchy, 101
remainder, Lagrange, 101
right-handed system, 22
Rolle's theorem, 87
roots of unity, 41

saddlepoint, 117
scalar product, 25
second derivative test, 85
SHM, 174
– damped, 177
solid of revolution, 79
standard derivatives, 45
substitution
– algebraic, 64
– trigonometric, 66
surface, 106
surface area, 133

Taylor polynomial, 99
Taylor series, 102
Taylor's theorem, 99
total derivative, 112
total differential, 110
trial function, 181
triple integral, 136
triple scalar product, 28
triple vector product, 30

upper triangular form, 8

vector
– differentiation, 31
– free, 21
– position, 21
– resultant, 22
– unit, 23
vector product, 26
vector resolution, 22
volume integral, 130